About the Author

PETER MAAS grew up in Redding, Connecticut. He graduated from Duke University, and was a staff member of *Look*, *The Saturday Evening Post*, and *New York* magazine. He is the author of more than a dozen books, including *The Terrible Hours*, *Father and Son*, *Underboss*, and *The Valachi Papers*. He passed away in August 2001.

SERPICO

SERPICO

PETER MAAS

HARPER ● PERENNIAL
www.harperperennial.com

A hardcover edition was published in 1973 by Viking Press, an imprint of
the Penguin Group (USA) Inc.

HarperCollins books may be purchased for educational, business, or sales
promotional use. For information, please e-mail the Special Markets Department at SPsales@harpercollins.com.

First Perennial edition published 2005.

Library of Congress Cataloging-in-Publication Data
Maas, Peter.
    Serpico / Peter Maas.—1st Perennial ed.
        p. cm.
    Originally published: New York : Viking Press, 1973.
    ISBN 0-06-073818-9
        1. Serpico, Frank. 2. Police—New York (State)—New York—Biography.
            I. Title.
HV7911.S4M3 2005
363.2'092–dc22

                                                            2004057285

                    HB 02.15.2018

*This is in memory of
Madeleine;
and also for
Vincenzo
and
Maria Giovanna*

PART
ONE

chapter 1

It is a warm September afternoon in New York as I watch Frank Serpico, age thirty-five, the son of a Neapolitan shoemaker, walk with the help of a cane toward the entrance of a fashionable Manhattan hotel. The hostility of the hotel doorman, white-gloved and resplendent in a forest-green, brass-buttoned, epauleted uniform, is immediately evident. His nose, with crosshatched tiny red veins, sniffs disdainfully; his watery blue eyes grow suspicious at Serpico's approach. Clearly he does not like what he sees.

Serpico is a short, muscular man, with a shock of brown, curly hair that brushes his shoulders and a full beard. He is wearing leather sandals, a pullover shirt of coarse white linen with leg-of-mutton sleeves, a leather jerkin, and brown velour trousers with flared bottoms. The trousers are supported by a wide belt with a huge brass buckle that Serpico found in a flea market. Emblazoned on the buckle are the heads of two bearded gentlemen of historical note, Henry Wells and William Fargo. Between them are crossed American flags and underneath the legend SINCE 1852. On Serpico's right wrist there is a silver bracelet, and on his left a double strand of varicolored quartz love beads. His shirt is open almost to his waist, and suspended from a slender gold chain around his neck is a gold Winnie-the-Pooh. It was given to Serpico by a Swedish girl he met during a trip to Stockholm. One night he was reminiscing about his childhood and happened to mention that the Pooh stories had been his favorite book, and the next day the girl went out and bought the gold figurine for him.

Serpico has dressed with some care for this visit uptown. In Greenwich Village, where he lives, he would normally turn out in a striped T-shirt and a pair of faded jeans which he himself, being handy with needle and thread, has repaired and patched from time to time. Still, the hotel doorman would like nothing better than to spot a small sign of hesitation on Serpico's part, a hint of indecision, anything to enable him to confront Serpico with an accusatory, "Can I help you?" But Frank Serpico has been through all this before; he knows exactly what the doorman is thinking, and he limps past him as if he did not exist.

I wonder what the doorman would do if he knew that inside the cane Serpico is leaning on—an eighteenth-century English cane of tightly wound stripped leather with a carved ivory cardinal's head for the knob—there is a twenty-nine-inch-long sword with a razor-sharp edge and point, or that beneath Serpico's jerkin, in a holster clipped to his belt on his left side, the butt facing forward, there nestles a big, loaded, fourteen-shot, 9-mm. Belgian-made Browning automatic pistol in well-oiled, working order. Serpico never goes out without the automatic; it is the reason why, even in the hottest weather, he always has on a jerkin or vest of some sort.

Serpico has just returned to the city after a two-week vacation in Nova Scotia, and since he had a doctor's appointment near the hotel, we arranged to meet there for a drink so I could hear about the trip. He orders a Bloody Mary and asks for a stalk of celery in it. The waitress, a pretty blonde, says with a touch of annoyance, "Celery? I never heard of that."

"You ought to try it," Serpico says. "It'll cure all your ills." He looks directly at the waitress as he speaks. Serpico is not, by any conventional standard, handsome. His nose, for example, is too large for his face and is bent slightly sideways, as if it once sustained a blow from which it has never recovered. But one's impression of him at any given moment is governed by his eyes. They are dark brown, and when he is angry, they smolder with rage. On the other hand, when he smiles, as he does now, they dance instantly with their own inner amusement, the lines around them crinkling in concert. Together with the suggestive note in his voice, the effect on the girl is magical. She smiles back, blushing, and says, "Oh, wow! I guess I will."

Over drinks, Serpico speaks longingly of his trip to Nova Scotia, of the brilliantly crisp days, the marvelous, silent nights. He had gone north while the word was carefully passed that he was headed south, even to the extent of purchasing an airline ticket to Florida. He had driven to Nova Scotia in his Land Cruiser alone save for his English sheepdog, Alfie, specifically because of a death threat on his life, but also to get away from the city for the first time in more than a year, to reflect on a series of personal crises, past and present, and to think about his future.

Except for a two-day stay with a farmer he encountered on the road, Serpico recalls, he spent his time driving leisurely along the coast, stopping occasionally to fish or to walk on the beach to exercise his left leg, which was still weak from a severe attack of phlebitis, a painful and sometimes dangerous inflammation of the veins that had first put him in a wheelchair, then left him with the cane.

At night he usually camped out. Serpico carried a large plywood board with attachable supports in the back of the Land Cruiser, and when he spotted a suitable site, he would set up the board so that it extended out through the rear double doors, place a sleeping bag on it, and rig a tentlike tarpaulin overhead in case it rained. Then he would build a fire, feed Alfie, and cook himself a steak he had bought or a fish he had caught during the day. After coffee, he would bed down in the sleeping bag and by lamplight read one of the books he had brought with him. He found *My Life,* by the great modern dancer Isadora Duncan, particularly appealing. Serpico's favorite diversions are the opera and ballet, and the Duncan autobiography had been given to him by a dancer he dated for a while. What he liked most about the book was Miss Duncan's fierce independence and determination to do, as he puts it, "her thing."

There was one incident on the trip that had a special significance for Serpico. He had arrived in Boston after midnight, and decided to sleep until dawn in a parking area near the South Boston waterfront. Around three A.M. he was awakened by Alfie's barking and a flashlight shining in his face. The flashlight, he quickly discovered, was being held by a Boston policeman, and Serpico braced himself for a series of sharply worded question about who he was and what he was doing there, perhaps even a search of the Land Cruiser for narcotics. But all the policeman did was warn him that there had recently been a number of muggings in the neighborhood. "It really made me feel good," he says now, recounting the incident in the hotel bar. "Here was a cop doing what he's supposed to do—helping people, not hassling them."

Serpico, who drinks sparingly, refuses a second Bloody Mary, and I decide to accompany him back downtown. On the way, he lights up one of the three or four cigars he smokes each day. He gets his cigars in a little shop near the garment center, hand-rolled on the spot for him. He has scouted places like this all over the city to satisfy his wants. In an era of plastic, prepackaged foods, for instance, he prizes the real thing and will travel extraordinary distances to obtain favored delicacies—to an obscure street in Brooklyn for a freshly made, spicy Polish sausage called kielbasa, still farther out in Brooklyn for the newly ground Turkish coffee he savors periodically, to an Italian butcher off Ninth Avenue in Manhattan who makes his own Genoa salami, to a cheese store on the Lower East Side for mozzarella so fresh that the milk spurts from it at the touch of a knife.

Serpico plans to take in a movie after dinner, and as we bounce along in his Land Cruiser he considers which girl to invite. The possibilities are enough to send an avid reader of *Playboy* magazine into a frenzy, and three of them immediately come to mind. One is his steadiest girlfriend, a carefree airline stewardess who looks as if she just walked in off the southern California beach where she was raised on sun and surfing.

Another is a somewhat tense, twenty-six-year-old, blond, hundred-dollar call girl with stupendous breasts. Serpico met her during the summer when he temporarily vacated his own place—for the same reason that later sent him to Nova Scotia—and moved into an apartment that a friend lent him. His path crossed hers one afternoon while he was walking Alfie around the block, and she was out with her toy poodle. About four o'clock the next morning, Serpico was still up, listening to his favorite opera, Puccini's *La Bohème,* and he decided to take Alfie out for another stroll. As he turned the corner, he came upon the girl trying to hail a taxi. He asked her where she was going at such an hour. The girl replied that she could not sleep and was going downtown to talk to some friends. Serpico said, "Why don't you talk to me? At least you'll save cab fare." They walked through the deserted city streets for a while before going back to the apartment, and from then on whenever the call girl wasn't professionally occupied she would see Serpico. Once she showed up just as he was preparing to take a bundle of dirty clothes to a Laundromat. She snatched it from him. "A man shouldn't have to wash his own laundry," she said. "That's woman's work."

There is also a volatile black model whom Serpico has been dating on and off for the past year. This summer, however, he has not seen much of her, and she has begun slipping angry notes under his door promising all kinds of mayhem for deserting her for "white chicks."

Serpico's normal residence is on Greenwich Village's west side, three blocks from the Hudson River docks. The building is a run-down, five-story affair; a water pipe burst over the front entrance some time ago, and the ceiling has yet to be repaired. His apartment is in the rear on the ground floor, at the end of a dingy hall. Serpico read about it in a newspaper ad offering a "garden apartment with wood-burning fireplace." The "garden" turned out to be a small triangular courtyard littered with broken bottles and tossed-away garbage, and an adjoining building effectively shut out the sun unless it was directly overhead. The apartment was equally small and dark, but it did have a fireplace that worked, so Serpico signed the lease. He cleaned up the courtyard, set out some potted plants, installed an outdoor lighting system, and fashioned a concrete basin which in warm weather usually contains some goldfish. A door off the courtyard leads to a sleeping alcove, barely large

enough for his bed, a ten-gear racing bike, and a bub-
bling glass case filled with tropical fish. The rest of the
apartment consists of a single, incredibly cluttered
room, twelve-by-fifteen, which manages a raffish élan,
especially in winter, with a fire blazing in the hearth.

In one corner, beneath a huge tapestry of a romantic nude surrounded by cherubs and a Tiffany
lamp suspended from the ceiling, there is a round,
white marble table which in a pinch can seat four. Like
most of the other furnishings in the room—a maple
chest, a small sofa, and a big armchair with a foot-
stool—they were picked up by Serpico in secondhand
shops around the city. In another corner is a kitch-
enette, partially hidden by a carved Indian screen,
where he expertly prepares such dishes as chicken
breasts with apricots. The horns of a Black Cape water
buffalo hang over the brick mantel, along with an
African mask, an ancient muzzle loader, and an old
pendulum clock from Copenhagen. Victorian brass
tongs, a shovel, and poker stand on one side of the
fireplace, an enormous copper kettle filled with wood
is on the other. Elsewhere a guitar leans against a
wall, a Ben Shahn peace poster above it; there is also
a television set, a stereo system, and a camel saddle
Serpico got in Morocco which can be pressed into
service as an extra chair. With Alfie's shaggy bulk
stretched across the floor—he has been trained as a
watchdog and begins growling at the slightest alien
sound—one must step about with considerable delicacy.

Stacks of records are under the twin windows facing the courtyard, and shelves in still another corner

of the room contain various mementos Serpico has brought back from his travels—Indian pottery and a bone chess set from Mexico, beer steins from Germany, a meerschaum pipe from Holland, and ebony fighting bull from Spain, a Venetian vase in which he keeps incense sticks. More shelves between the windows overflow with books, among them Thoreau's *Walden*, the collected poems of Yeats, Baudelaire's *Les Fleurs du mal, Steppenwolf,* Dalton Trumbo's *Johnny Got His Gun,* and a best-selling novel about cops by a Los Angeles police sergeant, *The New Centurions.*

After driving down from the hotel, Serpico finally finds a parking spot a block from his apartment. The space is just big enough for the Land Cruiser, and he backs into it with some satisfaction. He used to own a BMW sports coupé which became increasingly dented and battered on the city streets; now the other fellow has to worry.

The neighborhood is visually undistinguished, sandwiched as it is between picturesque, tree-lined Greenwich Village proper and the lofts, warehouses, and trucking garages that front the river. Rents are relatively cheap, and the population is basically young and anti-establishment. A good deal more marijuana is smoked than alcohol consumed. The men wear their hair shoulder-length and the women consider themselves liberated.

Serpico is a familiar figure there, and as we go along the sidewalk, he is continually greeted, like a politician making the rounds of his district. A jovial round-faced black man, who has developed a thriving business making artificial flowers for the uptown trade, yells at him, "Hey, dude, where you been?" But most of the people Serpico encounters—a pony-tailed brunette refinishing a table in front of an antique store, a law student in blue jeans, a couple who have just been given their first rock recording break—call him "Paco." It is Spanish for Frank, and Serpico acquired the nickname several years ago when he began taking frequent trips to Puerto Rico—among other reasons, to improve his knowledge of Spanish. He is a first-rate mimic and has an uncanny ability with foreign tongues. Besides Italian and Spanish, which he speaks fluently, he handles himself well in French and German, and even retains a smattering of the Japanese that he learned during his army service.

As we start to enter his building, a lanky young man in cowboy boots and a fringed frontier shirt hails Serpico and says, "Did you hear about that chick who got murdered on Jane Street last night?"

"No." Serpico drops his casual air and stares intently at the young man. "What time?" he asks.

"I don't know. Sometime last night. Why?"

"Was she shot?"

"I'm not sure, Paco. I think she was knifed, but I'm not sure."

Serpico tugs at his beard. "That's funny," he says. "You know, I thought I heard a shot around three o'clock, and I grabbed my gun and ran outside. But there were just a bunch of fags in front of that joint on the corner, and I couldn't get anything out of them."

For a long time none of his neighbors knew who Serpico was. The odd thing, considering their lifestyle and attitudes, is that now that they do it makes no difference that he is in fact Detective Third Grade Frank Serpico, shield number 761, of the New York Police Department.

If Serpico's friends in the Village were first stunned by this revelation, and then accepted it, his fellow cops were equally confounded by news of a different nature, and they still have not recovered. Serpico—this apparent hippie, womanizer, hedonist—had dared do the unheard-of, the unpardonable, in police circles. Having solemnly sworn to uphold the law, he elected to do just that, to enforce it against everybody—and not, in the grand tradition of even the most personally honest policeman, against everybody *except* other cops. He would not go along with the graft, the bribes, the shakedowns; and he refused to look the other way.

With that decision, Serpico became unique. He was the first officer in the history of the Police Department who not only reported corruption in its ranks, but voluntarily, on his own, stepped forward to testify about it in court. He did so after a lonely four-year odyssey in which he was repeatedly rebuffed in his efforts to get action from high police and political officials, continually risking discovery at any moment by the crooked cops he rubbed shoulders with every day, and finally, out of desperation, after he went to a newspaper with his story. This sparked a jumbled rush of related events, in which the administration of Mayor John V. Lindsay found itself hugely, and publicly, embarrassed: a Commission to Investigate Alleged Police Corruption was formed; the Police Commissioner abruptly resigned; there was a mass exodus of top-echelon police command; departmental organization and procedure underwent a revamping drastic beyond memory; in the wake of a new edict holding field commanders responsible for the actions of their men, precinct captains were banished to lesser posts; inspectors were reduced in rank, and lieutenants and sergeants were transferred in wholesale lots; a flurry of criminal indictments and departmental charges were announced; the head of a federal task force, launching his own investigation, declared that he had unearthed "large-scale payoffs to policemen" of up to twenty-five thousand dollars; then a parade of cops on the take, caught red-handed by commission investigators, attested in public hearings to widespread, deep-rooted corruption, with the graft in just one section of the Police Department amounting to four million dollars a year; and key municipal and police officials at last admitted under oath that, despite the specific allegations brought to them by Serpico, they had in fact done nothing.

As these events began to unfold, and Frank Serpico's role in them was made known, he became for many police officers a man to be feared—and destroyed.

Seven months before our meeting—at 10:42 on the night of February 3, 1971—the phone rang in the emergency room of Greenpoint Hospital in northwest Brooklyn. A petite black nurse, Ann Bennett, answered, listened for a moment, and called out, "A policeman got shot. They're coming in with him."

Some blocks away, a radio car raced up Metropolitan Avenue toward the hospital, its siren wailing, lights flashing. Serpico lay slumped in the rear seat, his face, beard, the upper part of the army fatigue jacket he was wearing covered with blood. Beside the driver, a second cop knelt back over the front seat, bracing his hand against Serpico's limp body to keep him from toppling over as the car swerved and cornered.

About twenty-five minutes had elapsed since Serpico, on a narcotics raid, wedged half in and half out the door of a heroin dealer in the Williamsburgh section of Brooklyn, had suddenly seen a gun poised directly at him in the dark alcove, perhaps eighteen inches away, and then simultaneously heard the roar as it went off and saw the flash—an enormous flash of colors, merging reds and oranges and yellows—and felt the searing heat in his head, as if a million white-hot needles had been plunged into it. He had been shot in the face.

Now, in the radio car, waves of weariness swept over him, and he struggled to remain conscious. He wondered how badly he had been hurt and vaguely heard the call over the radio that they were going to take him to Greenpoint and thought that maybe, if he could hang on until he got to the hospital, he would be all right. The cop in the front seat bending back toward him, holding him, said something that Serpico sensed were words of reassurance, but because of the siren he could not hear what they were, and for the rest of the trip the siren drowned out everything else.

The blood was beginning to cake on his face.

Before, lying on the floor after he had been shot, he had seen it spurting past his eyes, actually a geyser of blood splattering against a filthy wall inches away. He could still feel his mouth full of blood, as well as the sinus cavities around his nose. The sensation was the same that he had experienced when he accidentally got water in his mouth and nose while swimming. But mostly now, in the radio car, he concentrated on staying awake, clutching to consciousness like a talisman, fighting the insistent tiredness that pulled at him so seductively, thinking that if he could only keep his eyes open, somehow he would not die.

When the news was received at Police Headquarters that a cop had been gunned down and that the cop was Frank Serpico, the first tremors of official trepidation set in. Some six months earlier, Serpico had been the key witness in the perjury trial of a corrupt plainclothesman who had denied taking thousands of dollars in graft. The plainclothesman had been convicted and sent to prison, causing still others in related cases to plead guilty in hopes of lighter sentences. Thus the initial fear at headquarters was that Serpico had been shot by one or more of his fellow police officers. When this, in the hours immediately ahead, proved unfounded, there remained a second ugly question: Had the shooting happened in the normal course of duty, or had he been set up?

Similar thoughts swirled through Serpico's mind. The remembered encounters in the precinct houses. A cop he rather liked pleading with him, "Frank, why don't you take the money?" Another cop suddenly pulling a knife on him and snarling, "We know how to handle people like you." There was also that moment in court, recalled so many times that it came back to him now totally, without effort: sitting next to a gambler he had arrested, waiting for the arraignment, the gambler turning to him, pointing his trigger finger at Serpico and curling it, and quietly remarking, "Hey, you know they're going to do a job on you."

"Who?"

"Your own kind."

A few days earlier Serpico had personally closed down a Mafia-run betting operation in East Harlem which had been paying off the police. "What do you mean, my own kind?" Serpico asked. "The Italians?"

"No," the gambler replied. "Cops!"

Attendants were waiting with a stretcher outside the emergency entrance when the radio car wheeled into the hospital grounds. A new emergency complex was under construction, and Serpico was trundled on a circuitous route around it, down a dingy cream-colored corridor, past a waiting room and into the antiquated existing facility, a cramped treatment room which could hold no more than three patients at a time. As it happened, there was none there when Serpico arrived, and Miss Bennett, the nurse who had answered the phone, remembered looking down at his blood-drenched, inert form and saying to herself, "Oh, no, he's dead." Then she heard him mumbling something unintelligible and realized he was still alive, and she and another nurse quickly began to strip off his clothes—the padded army jacket, a leather vest, a heavy black woolen sweater, dungarees, and calf-high boots. Miss Bennett tried to get the gold Winnie-the-Pooh over his head and when she couldn't, she snipped the chain. Serpico heard the second nurse say, "Oh, look, somebody lost an earring." He knew the earring, a simple gold circlet, must be his, but when he tried to tell her he found he was unable to open his mouth.

A hospital smock was slipped on him, and the two nurses started cleaning the blood from his face and neck. Most of the flow had stopped by now, except for a watery, red-streaked dribble from his left ear. His left eye was beginning to blacken, and the left side of his face had ballooned to nearly twice its normal size and was paralyzed, giving his mouth an oddly twisted look. Then Miss Bennett saw the blood-encrusted hole where the bullet had gone in, just to the left of Serpico's nose right above the nostril.

A doctor was standing alongside her, and Serpico heard him say, "Don't touch it," and someone else said, "It's his head," and he wondered again how badly he had been hurt, trying to maintain his confidence and thinking to himself that at least he could see and hear. Lying there on the stretcher, ringed now by a blur of faces peering down at him, he felt utterly impotent, and he hated it. Above the faces were lights glaring down on him, and he suddenly recalled lights just like them, and the same sense of helplessness, the time he had been rushed to the hospital for an appendectomy when he was five years old.

As soon as his head wound was discovered, one of

the doctors severely pinched a muscle in Serpico's shoulder. He winced in response to the pain, a sign that his central nervous system was still functioning. Then he saw the needles going into his arms. One was to help stave off any swelling of his brain. Two others were to prevent convulsions.

Once more he was seized by an immense weariness, and he struggled to stay awake, fighting sleep, equating it with death and defeat. A doctor said to him, "Just take it easy. You're going to be OK." Vaguely, he became aware of blue uniforms mixed with all the white ones, and of a voice asking, "Are you Catholic?" Oh, no, Serpico thought, not *that*, not now. It had been years since he had been to mass or confession, but he must have nodded because somebody said, "Yes, yes, he's Catholic," and the next thing he knew a priest was leaning over him saying, "Can you hear me? Can you hear me? You don't have to answer," and proceeded to administer extreme unction, the last rites of the Roman Catholic Church.

When the priest was finished Serpico saw a uniformed cop standing nearby, notebook in hand. Suddenly there was an important message he wanted to communicate, and he motioned weakly to the cop. With tremendous effort, he managed to mumble out of the right side of his mouth, "Don't call my mother. Call my sister, my sister."

The cop looked nonplussed. "What's he saying?" he asked a nurse on the other side of the stretcher.

"My sister," Serpico mumbled again. "Not my mother." Maria Giovanna Serpico, a small frail woman, then seventy-two, not only had a history of heart trouble, but was recovering from a debilitating seige of the flu, and had been advised by her physician to remain indoors for at least another week of the New York winter unless an exceptionally warm and sunny day happened along.

More and more hospital personnel and police officials were crowding into the emergency complex, and as the babble of voices mounted, Frank Serpico was rolled past them into another room for X rays to determine whether immediate surgery was going to be necessary for the removal of the bullet inside his head. His condition was critical but apparently stable for the moment.

The X rays revealed that Serpico's survival thus far bordered on the miraculous. What had saved him, and could also have killed him on the spot, was that a bullet entering the body often does not take a logical course. Simple muscle contraction in response to the impact can be enough to change its direction, even through the soft tissue of the face, where Serpico was hit. If the bullet continued along its original line of fire, it would have struck the upper part of his spinal cord, paralyzing his arms and legs, bladder and bowels and respiratory system. If it had curved slightly upward, it would have smashed into the base of his brain and brought about instantaneous death.

As it was, the fact that Serpico was still alive could be measured in hairbreadth terms. The bullet—fired, it eventually turned out, by a .22-caliber target pistol—had not penetrated the skull. It had gone through his left maxillary sinus, one of the two cavities on either side of the nose. It then swerved slightly down and to the right, plowing into the back of his jawbone where it broke into fragments, two of them quite large, that were now lodged in the bony portion of his left ear. Although by all odds the jawbone should have been shattered as well, the angle of impact was such that this simply had not happened. On its erratic path, the bullet tore through some facial nerves, causing localized paralysis, but aside from the initial, spontaneous bleeding, no major veins had been touched. Serpico's escape, however, had been excruciatingly narrow. One of the fragments, after ricocheting off his jawbone, had stopped barely half a centimeter short of the big carotid artery running up the side of his neck, through which the heart pumps blood directly to the brain. Had it continued just that minuscule distance, he would have hemorrhaged to death in a matter of minutes on the dirty tenement landing where he had fallen.

Since the bullet had not gone into the brain, surgery was ruled out for the time being. Even so, Serpico's condition remained critical, and he was taken into another room in the emergency complex to be kept under close observation. The concussive effect alone of the shooting might have caused as yet undetected swelling and bleeding in the brain. More than one patient had come into a hospital emergency room with a head injury and died two or three hours later although no skull fracture was evident. There was, besides this, the persistent watery drainage from his left ear which was now recognized as cerebral spinal fluid. Normally this substance circulates through the closed system of the brain and spinal cord, and its ominous appearance meant that the cerebral membrane had been torn, allowing the fluid to leak out and exposing the brain to the possibility of a fatal infection.

Serpico felt no specific pain then, just a numbness in his head, and in the observation room he finally started to doze off, only to be awakened by a recurrent nightmare that he was suffocating. It was very difficult for him to breathe. His nose was almost completely blocked by caked blood, left untouched for fear of reopening the wound, and his mouth was dry from drawing air through his clenched teeth. Once he tried to raise himself and fell back, unable to do so, cursing inwardly. That was perhaps the worst part for him, his total dependence now upon others. He had always been geared to independence, self-reliance, and, beyond that, to having others rely on him. This as much as anything was what had attracted him to police work in the first place, the authority and the power that automatically went with it, and the ability to help and protect people. One of his favorite movies was *High Noon*, the brilliant portrayal of a beleaguered sheriff who almost single-handedly saves himself and his town from a band of killers. And his best memories of being a cop were of delivering the baby of a terrified mother in a crumbling ghetto room, of spending a day tracing down the whereabouts of a soldier on leave whose stolen duffel bag Serpico had recovered, of noticing a telltale wisp of smoke and evacuating an entire apartment house in the middle of the night moments before it went up in flames.

He was always privately scornful of cops who put in their eight hours on the job and let it go at that.

While he was waiting for his appointment to the Police Academy to come through, a captain had counseled him, "Never get involved in anything when you're not working."

"What do you mean?" Serpico asked.

"Kid, when you're off duty, you're off duty. Say you're driving home some night late and you see this guy breaking into a place. Well, keep driving. Otherwise, the next thing you know, they'll be asking you what the hell you were doing out at that hour."

Especially during his early days as a foot patrolman, Serpico, in civilian clothes after a four-to-midnight tour, would often seek out muggers on his own. Besides his talent for mimicry, he has an actor's ability with his body, and in a variety of guises—his favorite being that of an elderly man shuffling along over a cane, a big slouch hat concealing his features—he would go alone down dark and silent city streets in high crime areas, waiting for the attack to come, actually inviting it, his eyes probing each doorway for a sudden shadowy movement, his ears straining for the predatory footfall behind his back.

A girlfriend, lying next to him in bed one night, said, "You must be crazy doing things like that. What are you trying to prove?"

"I'm not trying to prove anything," Serpico replied. "I just want *them* to know how some poor slob feels when they jump him."

Once four young muggers jumped Serpico. He whirled, kicked the knife out of the hand of the leader, drew his revolver, and watched them freeze. He identified himself as a police officer and lined them up, hands against the side of a building, feet out. An awkward moment ensued. Despite the commotion, no help appeared. Then Serpico heard one of them mutter, "Shit, man, he can't take us all." He quickly stepped up to the knife-wielder and put the revolver to his head. "Any of you move," Serpico said, "and *he* gets it." Someone said, "Fuck *him*," and with that, the other three raced off, each in a different direction. Serpico held his fire, afraid that a stray bullet might hit an innocent bystander unseen in the darkness.

"Hey, man, you let them go. How 'bout me?" the remaining prisoner said, the revolver still pointed at his head.

"I didn't let them go," Serpico told him. "They escaped." Out of the corner of his eye, he saw the knife on the sidewalk and he was seized by a nearly overpowering rage. How many times had it been poised across somebody's throat, pressed into somebody's belly? Serpico shoved the revolver barrel against the mugger's skull. "You want to escape," he said, "go ahead."

Slowly the mugger slid down the wall, onto his hands and knees. "Don't shoot, man," he pleaded. "I ain't going nowhere."

At midnight, in the Greenpoint Hospital emergency complex where Serpico remained under close observation, Nurse Bennett was finishing her shift. Never had she seen so much top police brass, in and out of uniform, grimly milling about. Among them was the department's second-in-command, the First Deputy Commissioner, as well as the Deputy Commissioner for Press Relations, the Chief Inspector, three assistant chief inspectors, four deputy chief inspectors, and assorted inspectors and captains—just the kind of people Serpico had been trying to reach for so long, to no avail, in his battle against corruption in the Police Department that began with the fixing of traffic tickets and ended in the traffic of heroin.

As Miss Bennett left the complex, a man with a familiar face strode past her through the swinging doors. It was not until she got home that she realized he was the new Commissioner of Police, Patrick V. Murphy. Although Murphy had been in office for nearly six months, his appointment coming in the middle of the scandal rocking the department and the city, this would be the first time he had ever laid eyes on Frank Serpico.

chapter 2

When he was shot, Serpico was a member of a plainclothes detail in the Police Department. Plainclothesmen are actually patrolmen working, as the name indicates, out of uniform and on special assignment, usually in narcotics, prostitution, or gambling. While corruption in the police force was by no means limited to those on plainclothes duty, the temptations and opportunities it afforded for graft had always been especially high—in narcotics because of the huge profits at stake, and in prostitution and gambling not only because of the money, but because they were two areas of illegal activity that a large segment, if not a majority, of the public constantly demanded.

In theory, plainclothes work was a step up from

uniform patrol, and men of superior ability were supposedly selected for it. They were given specialized training in what to look for, how to conduct surveillances, the rules of evidence in making an arrest, and so on. But to most of Serpico's fellow trainees in plainclothes school, it was a waste of time; instead of attending classes, they could be out in the street collecting bribes.

Not all plainclothesmen, of course, were crooked. A few of them conveniently looked the other way or had their own quaint notions of how to handle the problem. One of the most personally honest police officers Serpico ever knew—a man who until the very last minute could not decide whether he was going to be a cop or a priest—was a plainclothesman. Serpico once discussed with him the widespread bribery and graft he was encountering, and asked his advice. His friend replied, "You've just got to set a good example so others can learn from it," and Serpico had no reason to doubt his sincerity.

Some honest policemen simply got out of plainclothes duty as soon as they discovered what it was like. This was the course chosen years before, for example, by Commissioner Murphy. The son of a cop, Murphy had impeccable credentials as a police officer. He was an aloof, soft-spoken, delicately featured man, with enough toughness and savvy and drive to have risen steadily through the ranks of the New York police force. He also served as the top police official first in Syracuse, then in Washington, D.C., and finally in Detroit before he was suddenly brought back to his native New York to stamp out corruption in its

thirty-two-thousand-man force—the nation's largest—
and to restore public confidence in its operations. It
was then revealed that early in his career he had
served a three-week stint in plainclothes. When asked
about the exceptional brevity of his tour, a spokesman
for the Commissioner explained that Murphy's wife
had just given birth and he had requested a transfer
"because of the bad hours" the work entailed. But
Murphy himself later admitted, "It was obvious to me
immediately that I was not going to be comfortable in
this kind of assignment."

Frank Serpico stayed in plainclothes because at the
time he was picked for the job it was the only way that
a patrolman lacking influential contacts in the department could advance to detective rank, and more
than anything else Serpico wanted to be a detective.
But while he would not participate in the organized
payoffs, he found in the end he could not ignore them
either. Instead he tried to do something about a system that allowed corruption to flourish. And it was
this that angered so many police officers, and left
them baffled and bewildered. He had broken an unwritten code that in effect put policemen above the
law, that said a cop could not turn in other cops.

Perhaps it would have been easier for them if Serpico fitted a recognizable puritanical mold. But he
dressed like a hippie and sported a beard and long
hair, and he lived in a bachelor pad in Greenwich Village doing, in their minds, God knows what. In the
suburban tract houses with tiny, neatly trimmed lawns
where most of the city's policemen lived, in the sa-
loons where they gathered, in the precinct houses and

radio cars, Serpico became the prime topic of conversation. One frequently repeated rumor about him held that he was "part spic and part Ethiopian, and speaks a funny sort of Italian," as if this, somehow, explained everything.

Around eleven o'clock on the night of February 3, right after Serpico had been brought into the emergency room at Greenpoint, his mother—in a cotton nightgown and robe which, like all of her clothes, she had made herself—switched off the television set. Her husband, Vincenzo, had retired almost an hour before, and she decided to join him in bed, although she did not find the prospect of trying to sleep very inviting after having been confined indoors for more than two weeks because of the flu.

Maria Giovanna Serpico was a doughty, voluble lady, barely five feet tall. Of her four children—three boys and a girl—Frank, the youngest, was the only one not yet married. On Sundays, at either her own home or her daughter's, she traditionally presided over a family dinner that featured staggering amounts of food, and under the lash of her cry, "What's the matter, you no like?" an unwary visitor was rendered helpless as serving after serving arrived, each a meal in itself, accompanied by bottles of the homemade wine that Frank's father fermented every year in his cellar. Frank attended these marathon affairs perhaps once a month, usually bringing along a friend, to whom Mrs. Serpico inevitably exclaimed at some point, "Eh, I bet you find the time to call *your* mother to see if she's still alive."

Vincenzo Serpico was as reserved as his wife was outspoken. He was a small man physically, but he carried himself with impressive dignity, and Frank always listened respectfully when he talked. He had worked from the age of nine, when he had been apprenticed to a shoemaker in Italy, and although they were reared in different worlds, Frank had a profound feeling for his father as a man of independent spirit, devoid of pretense, a master craftsman who took great pride in his work, who insisted on repairing shoes painstakingly by hand, and who, in the event a customer complained about the time this required, quietly advised him to go elsewhere.

Nothing in their lives had prepared these simple, decent people for the trauma they experienced the night their son was gunned down. But their essential nature and what they stood for, whether or not it was directly responsible for his lying helpless in Greenpoint Hospital, in very large measure had determined the kind of police officer Frank Serpico was. Apologists for the Mafia in this country, when they admit to its existence at all, like to point to the hostile environment Italian immigrants faced in America, to the inevitability, indeed the necessity, of banding together and falling back on old ways to survive, to the myth, for instance, that a given racketeer could have achieved equal success as a respected businessman if he were not ethnically cursed and denied that chance. But Vincenzo and Maria Giovanna Serpico had chosen another way and provided a very different kind of legacy, values that transcended any ethnic consideration—frugality, integrity, independence, self-respect.

When Frank was thirteen, he started shining shoes in his father's shop on Sunday morning after church, and during the week after school he worked for a fruit-and-vegetable street vendor. He was required to turn over his earnings to his mother, who gave him back a minimal allowance to spend as he wished. At first he was pleased with the idea of contributing, as he thought, to the household treasury, but later he began to resent being made to do it. Then, when he was seventeen and about to go into the army, his mother handed him a savings-account book, opened in his name four years before, with all the money and accrued interest in it. From that time on he never failed to put a portion of his wages into the account, and in the era of credit cards he paid cash for every purchase he made.

Vincenzo and Maria Giovanna Serpico had been married in a village not far from Naples, where they both grew up, although she was actually born in Ashtabula, Ohio. Her father had come to the United States in the 1890s and worked in the coal mines of West Virginia before moving on to a job in an iron-ore refinery in Ashtabula. But when Maria Giovanna was two, her father, having put aside enough money to become his own man, returned to Italy and purchased a house and some farmland, on which he planted walnut, prune, and fig trees, and began raising tomatoes and other produce, along with flax. One of her earliest memories was of her mother patiently manipulating a fork-and-paddle device to weave cloth from the fibrous plant; some of the linen sheets and pillowcases Maria Giovanna still used were similarly fashioned nearly half a century ago as part of her dowry.

Frank always delighted in hearing about the courtship of his parents, how his father fell in love with his mother while measuring her for a pair of shoes and the next day sent a boy with a message asking if he could see her. She replied that she needed a couple of weeks to think it over, but eventually she agreed to meet him in a local café owned by her aunt. After several such trysts, Maria Giovanna's mother got wind of what was going on and summoned Vincenzo. She told him she was not interested in having him hanging around "warming a chair," and demanded a timetable of his intentions. He explained that he was just starting out on his own, and it would be at least a year before he could assume the responsibilities of a wife and family.

Maria Giovanna had already had a number of suitors, all of whom had fallen by the wayside when her father refused to meet certain dowry conditions, and her betrothal to Vincenzo very nearly foundered for the same reason. Vincenzo's mother stepped in to handle the negotiations on his behalf, and among her requests was a house for the young couple to settle in. Maria Giovanna's father promptly turned this down, and for months the wedding was off. But Vincenzo at last took matters into his own hands, declaring that house or not he loved her, and they were married on August 27, 1925.

Two years later they decided to go to America. With Vincenzo's shoemaking talents, everyone agreed, he was certain to make a lot of money. By then the United

States had passed its restrictive immigration-quota laws, but Maria Giovanna was already an American citizen. At the time they were scheduled to sail, however, Vincenzo's papers still had not been properly processed. The Serpicos were advised that he would receive priority status if she were physically in the country, so it was decided that she would go on ahead. The plan was for an uncle who lived in Ashtabula to be at the boat and take her back with him to Ohio to await Vincenzo's arrival.

The trip was a series of horrors. Maria Giovanna was seven months pregnant when she left Naples, and in a rough crossing she gave birth prematurely. She landed in New York on a snowy December day in 1927, hemorrhaging badly, speaking no English, and with no one to meet her; the uncle who was supposed to have been on hand had become ill at the last moment. Authorities rushed her and the baby to a city hospital, but the baby, a boy, died. She remained in a charity ward for several days until distant family relations in Brooklyn finally tracked her down. Incredibly undaunted, and expecting Vincenzo to join her shortly, she stayed in Brooklyn with them. In return for a bed, she did the family washing and sewing and also went to work for a fur manufacturer; but the sweatshop conditions, the heat and stench of the skins, and the filthy hair-filled air proved more than she could bear, and she got another job on the production line of the Fanny Farmer candy company, rolling lollipops and inserting sticks in them. No saving was too small for her, and to avoid paying a nickel on the bus she walked two miles to and from work each day.

A year passed before she saw Vincenzo again, for even though his papers at the American end were in order, his passage was delayed by a Fascist edict that made it difficult for Italian nationals to emigrate. After they were reunited, they moved into a two-room, cold-water apartment in Brooklyn, and she continued at the candy factory while he looked for a job. When he found employment at last in a shoe-repair store, the cost to his pride was immense; the only opening was shining shoes for fifteen dollars a week. But within six months the regular repairman walked out following a dispute with the store's owner concerning the length of his lunch break, and Vincenzo took over for him. He remained there for six years, his weekly paycheck gradually rising to thirty-nine dollars, until the owner, claiming that profits had dropped off, cut his salary by ten percent.

The indignity of having to accept a reduction in wages was too much for him. Maria Giovanna knew this, knew what it meant to him as a man, and, although they had three children by then, unhesitatingly urged him to go into business for himself. Together they found a suitable shop in the neighborhood and all of their resources went into leasing it, buying materials, contracting for equipment. He barely survived the first few months, but slowly his trade picked up, and by the following year, when Frank was born, he had hired an assistant, and three years later still another one when he moved into larger quarters across the street.

Vincenzo Serpico eventually saved enough to purchase a modest, three-story brick house in the Bedford-Stuyvesant section of Brooklyn. For a while it lacked central heating, and Frank could recall when the only warmth in winter came from a coal stove in the kitchen. He also remembered how in the evening his mother would ready his bed with hot bricks wrapped in burlap, and how in the morning he would huddle under the covers to avoid the awesome chill of the toilet seat.

Although Frank inherited much of his mother's quick assertiveness, his father was the strong figure in his life, the one he consciously patterned himself after, and as a boy Frank loved to sit with him on Sunday afternoons in a greenhouse he had built in the backyard while he puttered among his plants, puffing a fearsome black cheroot and listening to the opera on the radio. The precarious state of the family's finances was over, but even so, the days when every cent had to be stretched, when nothing went to waste, were never forgotten. On one of these Sunday afternoons, when Frank was ten, he burst into the greenhouse crying because some kids had taunted him for wearing hand-me-down clothes. His father listened, and then told him a story about a prince who disguised himself in rags to see what the citizens of his kingdom were really like, only to be chased away by them. When the prince returned the next day dressed in full regalia, the townspeople who had hooted at him bowed and scraped, but he sent them packing, telling them they ought to be ashamed of themselves, that he was still the same person. "So you see," his father concluded, "it's not important how a man looks. It's what he is inside that counts."

The story, even the incident that triggered it, was banal perhaps, the sort of thing a father would seize upon to comfort a momentarily distraught son. But for Frank Serpico it was the most vivid memory of his childhood.

By 1971, the Serpicos could view their life with some satisfaction. Their eldest son, Pasquale, had a small grocery store that seemed to be doing well; their daughter, Tina, was married to an electrical engineer; another son, Salvatore, was running the shoe shop; and Frank was a policeman.

Mr. Serpico had stopped working two years before, at the age of seventy. He had always loved the earth, and now, able to indulge himself at last, he had acquired an acre of land in Long Island, where he spent the summer months cultivating vegetables with the same meticulous care he had given to the shaping and stitching of shoes. Once he telephoned Frank and mentioned casually something about a prize seventy-one-pound pumpkin from his garden. Frank knew instantly that the pumpkin was the real reason for the call, that his father wanted him to drive out to look at it, but that he would never suggest it without an excuse. So Frank said, "Seventy-one pounds! I don't believe it."

"No, it's true," his father replied delightedly. "You can come out and see for yourself."

Despite the country property, Serpico's parents refused to desert the Bedford-Stuyvesant house they had lived in for three decades. At the time they bought it the neighborhood had been a blend of Italians, Irish, and Jews. Since then there had been a drastic ethnic change, and the population was now mostly black and Puerto Rican. Because of their age—and also because of the racial tensions that had polarized much of the city, and the high crime rate in the immediate area—even Frank suggested that they now might consider moving, but the house represented such a deep psychological commitment to them, the first thing they owned in America after years of sacrifice, their first real roots in a new land, that whenever the idea was broached, they would say, "Maybe next year."

The ground floor was rented to a Chinese laundry, and the top floor to a middle-aged black couple. The second-floor apartment in which Mr. and Mrs. Serpico lived consisted of an eat-in kitchen, a living room, two small bedrooms and a larger one, fronting on the street, where they slept.

The night of the shooting, after turning off the television set in the living room, Maria Giovanna checked to make sure that there were no lights on in the rest of the apartment. She was fanatical about lightbulbs burning unnecessarily. Frank constantly poked fun at her about this, and she once triumphantly retaliated by showing him a magazine advertisement that pictured a young couple dining by candlelight. "See," she said, "it's not only me."

"Hey, Mom," he replied, "that's because they're supposed to be in love."

"What do candles have to do with love?" she retorted.

Now, in the darkened apartment, she tried to sleep, but being cooped up for so long with the flu had left her restless. As she tossed and turned, her husband mumbled irritably. Then she thought that her fever had inexplicably flared up or, even worse, that she was the victim of some strange mental aberration. On the wall opposite her hung a portrait of her parents, her father with his great, drooping mustache, and her mother with her hair parted in the middle and drawn back tightly in plaits. The painting had been done from a photograph, and she had always had mixed feelings about it since the artist had arbitrarily provided her father with a tie, an item of apparel he detested.

Suddenly there seemed to be a weird glow coming from the portrait. She was certain she had seen it, yet when she closed her eyes and opened them again, it had vanished. Then, as she looked at the portrait once more, it reappeared, a ghostly, wavering incandescence that clearly, if momentarily, revealed the eyes of her parents staring back at her.

She was not a superstitious person, but the whole thing was so eerie that she could not help thinking it must be an omen, that something was going to happen. She got up, went over to the painting, and inspected it a little sheepishly. There was nothing. Turning back to the bed, she glanced out of one of the windows overlooking the street, and stopped short. A radio car was parked at the curb, and a policeman was standing next to it. A second officer was on the sidewalk. Now she knew what had caused the glare on the

portrait, as the beam from the flashlight he was holding swept past the window.

She saw him move toward the downstairs door,
and she was already hurrying into the living room
when the bell rang. She buzzed back and stepped out
on the landing, a slight figure, clutching her robe
around her, her wispy gray hair in disarray, peering
down the stairs through rimless glasses.

At the bottom of the stairs the cop with the flashlight looked up at her and said, "Are you Frank Serpico's wife?"

The fears she had been struggling to subdue
flooded through her. "No, he's my son," she said.
"What's the matter?"

"He's been shot."

She felt herself starting to sway, and she gripped the
banister. Oh, God, she remembered thinking, give me
the strength to endure this. Then she heard the cop
say, "Don't worry, it's just his arm. We'll take you to
the hospital. Are you alone?"

"No," she replied. "My husband is here. He's sleeping."

"Are there any brothers or sisters?"

If it's just his arm, she thought, why are they asking about brothers and sisters? But she could not
bring herself to say it. The nearest of her children was
her son Pasquale, and she numbly gave his address.

"We'll get him," the cop said, "and come back for
you."

"Thank you," she whispered, and went back into
the apartment. In the bedroom she shook her husband. "Vincenzo, wake up," she said in Italian. He

groaned. "Wake up," she said. "It's Frank. He's been shot!"

Mr. Serpico sat straight up, fumbling for his glasses. "*What* are you saying?"

"The police, they were here. They went to get Pasquale, and they're coming back for us. Frank is in the hospital. They say it's just his arm. Vincenzo, I'm frightened."

Besides her recent illness, Maria Giovanna had been having problems with her heart for some time, and Vincenzo was concerned about the stress she was under and the cold night outside. "Giovannina," he said quietly—it was a diminutive he had used since they were married—"maybe it's better that you wait till morning. Pasquale and I will see about Frank."

"No," she said. "I'm going."

They dressed and then watched together by the window. When the radio car returned, they hurried downstairs and joined their eldest son in the backseat. The trip to the hospital was almost wordless.

His father had been rather proud when Frank became a policeman; it was a position of respect, an admirable transition into the American mainstream that he could never have hoped to achieve for himself, his own son moving past him, as he believed it should be. And now it had turned into a nightmare.

Frank's mother had not been pleased by his decision to be a cop, and during Frank's rookie days, when he still lived at home, she would wait up at night until he returned from a late-duty tour, terrified that something might happen to him. Gradually, however, she had overcome her fears, and although she was aware of his struggle against corruption in the department, the thought that he might be in danger from fellow police officers was beyond her comprehension. Then, in the radio car, on the way to the hospital, this possibility leaped so explosively in her mind that she believed for a moment that she had actually uttered it out loud, and she stared guiltily at the two patrolmen in front of her.

The torment of Frank's brother Pasquale, a gentle, reticent man, was even worse. Less than three months before, two officers had been dismissed from the force for trying to shake him down when he, like hundreds of other proprietors of neighborhood groceries and delicatessens throughout the city, had remained open on Sunday in a possible contravention of the state's crazy-quilt, so-called Sabbath laws. Frank had gathered the evidence against the two cops with a hidden wire recorder, and Pasquale could not help wondering if this, perhaps, was why he had been shot.

At Greenpoint Hospital, they were led to a bench in the emergency waiting room, still swarming with police officials, and when Mrs. Serpico realized that she was not being taken immediately to Frank's side, she collapsed in near hysteria. "My son, my son!" she began crying out. "Where is he? I want my son!"

As her husband tried to calm her, a doctor appeared and said that Serpico was being "worked on," but that they would be able to see him in a few minutes. If it's just his arm, she thought again, why are they taking so long? Before she could ask, however, the doctor was gone. Someone gave her a paper cup with water. She tried to sip it, but she could not get it down.

Then the priest who had administered the last rites to Serpico sat beside her. The unexpected presence of the priest was perhaps most frightening of all, so much so that it forced her to pull herself together. The priest, who spoke Italian, sought to console her. He said, without going into specifics, that her son had been gravely hurt, but that the doctors were very optimistic, and he was praying for Serpico's speedy recovery.

She was listening, as if in a trance, when she saw Commissioner Murphy approach. As the new head of the scandal-ridden Police Department, he had been getting a good deal of publicity, and she recognized him at once from newspaper photographs and from seeing him on television. He was the most important public personage she had ever met face-to-face, and she instinctively started to rise. "Please don't get up," Murphy said. He expressed his regrets for what had occurred and told her there was every indication her son was going to be all right. "If there is anything I can do for you, anything you want," he added, "don't hesitate to call me."

A nurse came and pointed to a door off the waiting room. "In there," she said. Mrs. Serpico had not realized that Frank was so close by, and she put a hand to her mouth, thinking to herself, My God, did he hear me crying?

Her first impulse, when she saw him, was to scream, but she managed to choke it back. Serpico was lying on his back, his right eye partially open, unfocused, the left one completely swollen shut. She could hear his labored breathing. A bandage had been placed

over the bullet's entry wound. Gauze, already tinged
with blood, had been loosely folded over his draining
ear, and she recalled that it looked as if a flower had
sprouted there. Two rubber tubes were connected to
a vein in the upper right side of his chest. One had
been used to measure his central venous pressure,
and was still in place in case he should require emergency treatment. Through the second tube he was
being fed an intravenous solution of glucose and
water as a further guard against shock and to prevent
dehydration. Dried blood was matted in his beard and
his hair and on his chest. She wanted desperately to
kiss him, but because of the tubes and bandages and
blood, she did not know where to do it. At last she
simply took his hand and kissed it and held it tightly.

Frank Serpico's father and his brother stood rigidly
silent next to her. Tears streamed down their faces.

When his mother gripped Frank's hand, he turned
his head and saw her. All he could think was that he
had not wanted her there, did not want her to see him
as he was. After a momentary soundless movement of
his lips, he whispered weakly out of the side of his
twisted mouth, "Don't worry, I'm OK."

They stayed with him for about an hour. Finally a doctor persuaded them to leave by convincing them that
what Frank needed now was to rest. The police brass
had departed as well, and at two-thirty that morning
he was taken to a surgical ward on the third floor.

Greenpoint was one of the city's oldest municipal
hospitals, and the ward reflected its age. It was a cav-
ernous room two stories high, full of exposed piping

and painted an unappetizing institutional green, with partitions dividing patients into groups of four—except for two cubicles, each barely large enough for a single bed, to the left and right a few feet past the desk of the supervising nurse.

Serpico was wheeled into the one on the left. He was placed on his side so that his draining ear faced up. His head was still numb. His nose remained clogged with dried blood, and his greatest discomfort came from having to suck air through his teeth. His mouth was terribly parched. He asked for a drink of water, but was told that he could not have it. At this delicate stage—hours after his cerebral membrane had been torn and with a bullet fragment nudging an artery to his brain—any gagging or, worse yet, vomiting could be fatal.

Assistant Chief Inspector Sydney Cooper got the call at home about two A.M. A thirty-one-year veteran on the force, Cooper was so accustomed to having the phone ring in the middle of the night that he could pick up the receiver, discuss whatever the problem was, and go back to sleep without missing a beat. This time, though, he sat bolt upright, shaken as he had not been in his entire police career. A reporter for *The New York Times* was at the other end of the line; he had been covering the burgeoning scandal triggered by Frank Serpico's revelations, and he wanted to talk about the shooting.

There was little Cooper could say, since this was the first he had heard about it. The irony was exquisite. Of the department's high command, Cooper was the only one Serpico trusted, the only one who wanted to pursue and expand an investigation into the corruption Serpico had reported—an investigation most of Cooper's colleagues tried to limit as much as possible in the hopes that the whole thing would blow over. He alone had befriended and backed Serpico, had seen him as a human being under enormous pressure, and not simply a pawn to be used or discarded to suit the department's pleasure. But in the panic that followed the shooting, no one in the department had bothered to tell him that Serpico lay critically wounded in the hospital.

Cooper drove immediately to Greenpoint. He was a big, balding man, and he liked to present himself to the world as a gruff, outspoken officer who barked orders at the top of his lungs. But as he stood in the darkened cubicle looking down at Serpico, he was overwhelmed with emotion. He knew how much Serpico wanted to be a detective, and finally he said, as steadily as he could, "Hey, you don't have to go to such extremes to get the shield." Then he, too, reached for Serpico's hand and held it.

Twenty years of police service separated Sydney Cooper and Frank Serpico, but they both had the same love-hate relationship with the police. For all of his carefully cultivated rough exterior, Cooper was an extraordinary and gifted man, accomplished enough as an artist to have been eligible for a Fullbright scholarship to study and paint in Italy for a year, and he had also earned a law degree. But he had not been able to bring himself to quit the police force. What made him stay was a special quality that every really good police officer seems to possess—an almost boyish delight in playing cops-and-robbers. "What the hell," Cooper once growled as he tried to explain it, "where else can a grown man have so much fun?"

But he was not an innocent either. He was perfectly aware of the ugly side of being a cop. During his own early days on the force, the attitude of his superiors about graft was, "Just don't get caught, fellows." When he achieved command of a precinct, however, Cooper began to crack down on the systematic shakedowns and bribes that had become a way of life among so many policemen. This was such a singular event, and Cooper was so successful at it, that he was eventually brought into headquarters to help combat departmental corruption on a citywide basis. But he soon found that the police hierarchy was far less interested in rooting out internal corruption than it was in hushing it up. He would always recall a summit conference on the subject during which one official spoke of "traditional practices that disgrace the department," while everyone around the table nodded solemnly.

"For Christ's sake," Cooper bellowed, "you sound like a bunch of old ladies talking about the clap!"

Although he was put in charge of several different investigative units with impressive titles, their budgetary and manpower restrictions made it virtually impossible for him to mount an effective campaign against corruption. Still, he did the best he could, and there was hardly a men's-room wall in precinct stations around the city that did not have an obscenity addressed to him. As a Jew in the Irish-dominated department, he knew what it was like to be an outsider, and he took the abuse philosophically. When he was asked about it once, he replied, "Well, I have this nightmare, you see, where I'm on Fifth Avenue watching the Saint Patrick's Day parade, and I have a coronary, and nine thousand cops march happily over my body."

Despite his reputation as an aggressive fighter against corruption, or because of it, Cooper did not know that Serpico had made charges of widespread police payoffs until at least two years after this had been brought to the attention of a handful of other high officials in the department. He finally learned about them when he was appointed commander of all patrolmen and plainclothesmen in the Bronx. It was there that Serpico's allegations were centered and seemed on the verge of exploding into headlines. Assigning Cooper to the Bronx was an ideal solution for the department. Whatever happened, it could always claim that it had dispatched the best man available, Sydney Cooper—hard-nosed, fair, professional, above reproach.

As Cooper now grasped Serpico's hand in Greenpoint Hospital, he remembered the first time they had met. Serpico had been in a tense, defiant mood, convinced that nothing would come of his charges, that at best a couple of patrolmen would be thrown to the wolves and that he would wind up an outcast, his career finished. Serpico angrily told Cooper that he would not testify in any court cases under these circumstances. "Why should I?" he said. "Who gives a fuck about me? I don't have a friend in the department."

"Friends!" Cooper roared. "Don't talk to me about friends. Don't give me that bullshit!" His face reddened. He slammed his fist on the desk, and stood up. "You know what I've been doing," he shouted. "I'm supposed to be the bogeyman around here. But I care about this goddamn department. You sit there bellyaching that you're alone. You don't think I'm alone? I've put cops away, and not just little ones. You think I could do it if I worried what they say about me? Fuck them! I'm not in any cliques. I don't have any friends in the department either, and I'm not looking for any."

It was a virtuoso performance, and Serpico could not help smiling. "OK, Chief," he had said, "I'll tell you what. *I'll* be your friend."

When Cooper left Greenpoint, he drove directly to the 92nd Precinct, where the shooting had occurred and the investigation into it was being conducted. He was as depressed as he had ever been, and once he was inside the station house he gave in to his worst fears and ordered a round-the-clock guard for Serpico.

Serpico had one more visitor that night. Around four A.M. the black model he had been dating was getting ready to go to sleep in her Greenwich Village apartment when she heard a radio newscast about Frank's having been shot and taken to Greenpoint Hospital. She threw on some clothes and ran out to find a cab.

At the hospital she announced that she was Frank Serpico's wife and demanded to be taken to him. Her manner was so authoritative that nobody questioned her. In the cubicle she stared at him and said, "Oh, baby, what have they done to you?" When she touched him lightly on the arm, he opened his eyes and motioned her to bend closer, whispering, "Water."

She went out and asked a nurse for a glass of water for him. The nurse explained that he could not have it. "Well, get me something," she snapped. "The man's lips are so dry they're starting to split." There was just the right degree of sharpness in her voice, and the startled nurse quickly handed her a pan of water and a washcloth.

She returned to the cubicle and sat by his bed, gently pressing the wet cloth to his mouth every few minutes. It was the first definite sensation of physical relief that he had experienced since the shooting; and as an occasional drop of the deliciously cool water rolled across his tongue, the idea that he was going to come out of this alive finally took hold. At about five-thirty Serpico fell asleep.

Shortly after eight A.M. a nurse awakened him. When she asked him how he was, he said, "OK. I have some pain in the left side of my face." It was slightly easier for him to speak, but he still had to mumble out of the corner of his mouth.

The nurse began sponging off the dried blood on his face and neck and in his beard that had been missed in the emergency room. "I have to pretty you up," she said. "You want to look nice for the Mayor, don't you?"

"Who?"

"The Mayor. He's downstairs, and he'll be up here any minute."

"What the hell does he want?"

The nurse laughed. "That's no way to talk, a big man like the Mayor coming to see you."

Alerted by a press aide that he was going to Greenpoint, a horde of newspaper, television, and radio reporters was already at the hospital when John Lindsay arrived, and as he was ushered to the third floor, they trooped along. Outside the ward, doctors told Lindsay that while Serpico's condition was stable at present, he was a "potential candidate for complications in the brain." The tall, photogenic Mayor listened solemnly as television crews began filming, and it was only with the greatest difficulty that hospital officials prevented the cameras from following him into the ward itself.

Serpico saw him come into the cubicle, but made no effort to acknowledge his presence. There was an awkward pause before Lindsay spoke. "Officer Serpico," he said, "I realize you're very weak, but I wanted to tell you that you are a very brave man, and that all New Yorkers are proud of you."

All Serpico felt was disgust. Nearly four years ago he had gone to one of the Mayor's closest aides, expecting that he would subsequently meet with Lindsay himself, and had detailed incident after incident that he had personally witnessed which demonstrated wholesale corruption in the Police Department. Nothing had happened. I had to get shot, Serpico thought, before he could find the time to see me.

The visit last possibly a minute. Afterward in the

corridor outside the ward Lindsay told the waiting reporters and cameramen, "He is a very brave man. He deserves the highest praise."

A reporter wanted to know if Serpico had said anything.

"No," the Mayor said, "he was too weak to respond."

chapter 3

Early in the morning of February 4, immediately after the Mayor's visit, the Police Department announced that Serpico's assailant had been apprehended. He was a twenty-four-year-old heroin addict and pusher named Edgar (Mambo) Echevaria. During the course of his capture Echevaria had been wounded in the stomach and had in fact been put in the fourth-floor surgical ward at Greenpoint, directly above Serpico. Echevaria's previous arrest record included possession of narcotics and felonious assault with a knife.

There was, the police said, no apparent connection between the shooting and Serpico's anticorruption activities, although an investigation *was* under way. Meanwhile, as the news spread through the department about what had happened, crudely scrawled notices appeared on bulletin boards in a number of precinct houses sardonically asking for contributions to hire a lawyer to defend "the guy who shot Serpico," and to pay for lessons to teach him to shoot better.

According to the police version of the shooting given to the press at the time, Serpico, backed by a team of plainclothes partners, had attempted to enter an apartment Echevaria was in after first identifying himself as a police officer. His path, however, was blocked by a chain lock. Two shots were then fired from within the apartment, one of which struck Serpico, and his partners fired back at once, hitting Echevaria in the arm. After he had fled out of the rear of the apartment, the police added, Echevaria was captured by other officers in a second shootout a couple of blocks away.

This was not quite how Serpico would remember it. He was not aware, however, of the details of the story as it was released that morning, and even if he had been, he was too sick to care. He was now experiencing terrible pain in his head, which came in pulsating waves, although this was actually an encouraging sign as far as nerve damage was concerned.

The sinister, blood-streaked fluid continued to ooze from his ear. While gauze had been loosely placed over the ear, it was there to absorb the drainage, not to hinder it. In a situation like this none of the normal procedures in bleeding applied; there was no attempt, for example, to staunch the flow because it would simply back up and accumulate between the cerebral membrane and the brain or inside the brain itself. The great danger Serpico faced was infection. The outer ear swarms with germs, and until the rip in his cerebral membrane mended all that could be done was to give him massive infusions of antibiotics. If any germs were drug-resistant, cerebral meningitis would be the likely result, and he would die.

Several medical conferences were held about Serpico that morning. Greenpoint did not have a neurosurgery department, but it had a neurosurgeon on call who was on the staff of Brooklyn Jewish Hospital, and it was finally decided to transfer Serpico there. The ostensible reason was that any surgery would be elective, rather than urgent, and that facilities at Brooklyn Jewish were better. But it was also a private institution; Serpico would be able to have his own room and nurses, and, as one of the doctors involved in the discussions put it, there was a feeling in the air that if anything went wrong, the Police Department did not want to leave itself open to charges that it had "short-changed" him.

In the early afternoon a police ambulance moved Serpico to Brooklyn Jewish. He was placed in a room on the ninth floor, and a uniformed cop took up his post outside the door. When the cop went off duty, a nurse heard him say to his relief, "Stay out of the room. Don't talk to the guy."

"Yeah, how come?"

"Didn't they tell you at the precinct?"

"No, they just told me to come over here."

"Well, I'm telling you. The word is don't talk to him. He's no good."

In a way Serpico had come home again. Brooklyn Jewish Hospital was in Bedford-Stuyvesant, only a block from the house where he had grown up. It was the same hospital that he had been rushed to for his emergency appendectomy, and it was the one he had gone to when at the age of fourteen he accidentally shot himself. Frank had always been fascinated by firearms, and as a boy he maintained status in the neighborhood gang he belonged to with his expertise in making zip guns. They were all the rage then in gangs throughout the city, crude contraptions in which tightly wound rubber bands powered a .22 bullet through a piece of a car radio aerial that served as the barrel. The bullet came out of the aerial with a tumbling action, instead of spinning, and at short range could inflict an ugly wound. Serpico had just finished putting one together in the cellar and was about to try it out when it suddenly went off and tore a bloody furrow down his arm. He was unable to stop the bleeding and, afraid that his mother would discover what he had done, he trudged to the hospital on his own. When he vaguely explained that he had been fooling around and that a bullet caused the wound, a nurse called the police. Two patrolmen arrived as he was being sewn up. "OK, kid," one of them said. "What happened to the gun?"

Serpico was petrified. His interest in guns was equaled only by his awe of the police, and he saw himself in handcuffs being taken down a long corridor, a cell door slamming behind him. "There wasn't any gun," he finally said. "I found this bullet and I put it in a vise to take off the tip, and it just exploded on me."

"Just exploded, huh?" the cop said.

"Yes, sir."

"Where do you go to school?"

"Saint Francis Prep," Frank replied. It was a well-known Catholic school in Brooklyn that his parents were finally able to afford to send him to, after his brothers and sister had attended public high school.

"Anybody who goes to Saint Francis," the cop said, "ought to know better." The two officers retired to a corner of the room until Serpico's arm had been bandaged. Then the one who had done all the talking said, "OK, we'll let you go this time, but don't let us catch you in something like this again."

Frank was sure that the two patrolmen knew he had concocted the story about the bullet, but had taken the trouble to size him up and decided that he was not really a bad kid, and wanted to avoid getting him into more trouble. That impressed him greatly. The police were there not to enforce the law blindly, but to help people as well.

Although Frank belonged to a street gang in his youth, that was simply a matter of survival in the streets of Brooklyn; he had always wanted to be a cop. Later, when he became one, and other cops were suddenly his enemy, he tried to pinpoint the exact moment of his decision, but he never could. As a boy he never missed a radio episode of *Gang Busters*, in which justice relentlessly triumphed. He traded a scooter to a friend for his Cub Scout shirt, not because he wanted to be a scout, but because it was the same blue as a cop's shirt. A member of his gang recalls seeing Serpico, when he was eleven, go by, clinging to the back of a trolley, and calling to him to jump off to play, and Frank yelling back, "I can't! I'm on a case!" On his twelfth birthday his parents gave him a microscope, and the first thing he did with it was to examine a dead fly, searching for clues that might explain the circumstances of its demise. Other boys in his neighborhood joined the Police Athletic League so that they could be taken to watch the Brooklyn Dodgers for free; Serpico was bored by baseball and became a member only so that he could sport the PAL button.

But what he remembered most from his boyhood was a concept of the cop as "good"—and beyond this that a cop was the personification of authority and prestige and respect. When a cop came down the sidewalk and said, "Move along," one moved, and fast. If there was a family dispute in the neighborhood, the cop on the beat was automatically called in as the final arbiter. He would never forget the first arrest he had ever seen—a thief dashing down the block, a cop commandeering a car, poised on the running board with his gun out, leaping off and covering the crook as he cowered, hands up, against the wall. And he could recall, as a kid shining shoes in his father's shop on Sunday morning after church, seeing the enormous ruddy-faced, white-gloved figure in blue, whom even his father treated with great deference, looming over him in the chair, his revolver bulging under his coat; the cop never gave him a tip, but in the proximity of such power it never occurred to Serpico to expect one.

Perhaps he had expected too much. He would look back on it and think how ridiculously naive he had been, how absurd to have accorded almost demigod rank to a man simply because he wore a uniform and a badge, and carried a gun and a nightstick. Still, it was a long while before his idealism faded.

Serpico enlisted in the army at seventeen, instead of waiting to be drafted, and spent two uneventful years as an infantryman in Korea, the monotony broken by periods on the rifle range, which he loved, lessons in karate, which he became very good at, and furloughs in Tokyo. He met a Japanese girl during his first visit, whom he always stayed with when he went back, avoiding the usual GI haunts, learning the language, trying to soak up as much as he could of Japanese life. It was the same pattern he would eventually pursue when he was a cop, living in Greenwich Village instead of mixing socially with other policemen, refusing to join what in effect was their closed society. But that, too, came much later.

After his discharge from the army, he immersed himself in everything he could that smacked of police work. He was not eligible to take the required civil-service test until he was twenty, and could not be sworn in as a member of the force until he was twenty-one; even this latter regulation was academic since there was a two-year waiting period because of the backlog of applicants. So he enrolled in Brooklyn College, majoring in police science. The college offered courses both day and night, and a number of cops working toward degrees were also students there. He noticed that most of them tended to wear white socks, dark slacks, and navy blue raincoats, and he immediately adopted their dress. He became friendly with a young cop he met in one of his classes, and on several occasions they double-dated. He would watch with envy as the cop simply flashed his shield instead of paying whenever they got on a bus, or if they went to the movies the way they were promptly ushered into the theater. Once, sitting around having coffee at a girl's house, Serpico asked him if he could look at his shield. He managed to slip in another room with it, and he held it up to his chest and posed in front of a mirror, experiencing a new sense of identity, thrilling to it.

Except in such police-science classes as criminology and how to conduct investigations, Serpico was an indifferent student, and he finally cut back on his course load so that he could take a job with the William J. Burns International Detective Agency. But his first assignment did little to fulfill his fantasies of a private operative working against a backdrop of world intrigue, mysterious women, and slick confidence men. Most of the agency's business involved security work, and he was told to report for duty as a uniformed guard at the Schaefer brewery near the Brooklyn waterfront. At least when someone he met in class asked what he did, he could say, without going into detail, that he was with the Burns detective agency, but the thousands of cases of beer he had to stare at week in and week out were too much for even his imagination, and he finally quit.

He continued to prepare for his police-entrance examinations, devouring every pamphlet and bulletin the department issued, one of which he taped to the wall above his bed. It stated, among other things, that a police officer exists only to serve his community; that he is out "pounding the pavement" while most citizens are asleep; that he "holds the broken body of a child in his arms and then seeks the calloused hit-and-run driver" who did the deed; that he "suffers intensely when one of his fellow officers is killed or injured" in the line of duty; and that he "suffers even more when one of his number turns black sheep and brings shame to all the others."

On the first available date following his twentieth birthday, Serpico submitted proof that he had completed a high school education, took the competitive test for prospective police officers, finished in the top ten percent, and was placed on the list for the Police Academy.

While he waited for his appointment to come through, he began to consider what he might like to specialize in once he was on the force, and decided that working with street kids on the youth squad would be rewarding. To build up some experience in this area, he applied for a job with the Youth Board, a division of the Mayor's office, which concentrated on keeping peace among the various juvenile gangs that were at each other's throats all over the city in the 1950s. A youth worker theoretically had to have a college degree, but the board was shorthanded at the time, and Serpico argued successfully that his own background of growing up in a gang environment made up for his lack of formal schooling. He was assigned to an aggressive, predominantly Italian gang in Queens called the Corona Dukes. His predecessor, an object of scorn because of his prissy manner, had left in near hysteria after one of the gang defecated in his hat. The first time Serpico met with the gang, several girls were summoned, and he was asked, "Hey, want to get laid? Take your pick."

Serpico did not respond with the lecture they expected on the evils of promiscuous sex. "Look, I don't need your help," he said amiably. "I can get all the stuff I want on my own." The next day he arrived with a supply of condoms and passed them out. "If you're going to screw around," he said, "you might as well not get yourself a dose." His techniques, if unorthodox, were effective, and during the seven months he worked with the gang, not one fight—or rumble, as it was called—occurred. When he got to be a cop, however, nothing came of his application for duty on the youth squad. Eventually he learned that it was a sought-after assignment because of the easy graft it brought from operators of billiard parlors, bowling alleys, dance halls, and other places that juveniles tended to gather where there were laws restricting the age of clients, the hours they could be on the premises, or the beverages they could be sold.

Serpico was sworn in as a probationary patrolman on September 11, 1959. Before taking the oath, he and the other recruits were given their shields. Some of the shields were new, but most of them had been previously worn by police officers now retired or promoted to sergeant. A few of them had only three-digit numbers, and others up to five. If someone lodged a complaint against a police officer, the most common method of identifying him was through his shield number, and Serpico heard a recruit behind him whisper, "I hope I don't get one of them low jobs. They're too easy for people to remember." Serpico was given number 19076. The shield was badly tarnished, and as he pinned it on he was already wondering if he could polish it enough to get the tarnish off. Then he was told that the shield could be replaced, and after the ceremony was over he raced out to have it done.

There were three hundred recruits in his class, divided into companies of thirty men each. During their orientation period, after they had received their gray probationary uniforms and .38-caliber police service revolvers, they discussed among themselves what they had done before entering the academy. When Serpico said he had been with the Youth Board, one recruit seemed surprised. "Hey, that was like you were a social worker or something!"

"Yeah," Serpico replied. "So what?"

"Don't get me wrong. I mean, it's just funny you being a social worker and all, and now you're a cop."

"What were you doing?" Serpico said.

"I was in construction. The pay was OK, but it wasn't an all-year thing. Like, if the weather was bad, you don't work, and maybe they stop building for a while, and jobs are tough if you don't have any seniority."

Except for a recruit who had been a detective in a department store, nobody seemed interested in being a police officer in terms of what he could do, what he could accomplish. Serpico remembered one recruit who said he had trained to be a barber but didn't really like the work and figured, what the hell, he would take the police test and see if he could pass it. Another had been a truck driver, but the work wasn't steady enough. A third, black, said it was a lot better than pushing a cart in the garment center. The emphasis was always on the economic and social advantages. It was clean, respectable work. The pay was the most one could get for the least education. Even though they were all in their twenties, the talk constantly revolved around job security, the free meals and discounts they could expect on the job, the pensions they would get. Serpico's first formal class at the Police Academy reflected the mood perfectly—it was on how to go on sick leave.

His recruit period lasted four months, divided between classroom instruction, pistol practice, physical training, and fieldwork. The physical-training instructor, a barrel-chested, bald-headed man known as "Mr. Clean," continually reminded his sweating charges of the financial advantage of being fit while making an arrest. "You don't want somebody doing chin-ups on your uniform," he would shout. "Remember, you paid for it!" Classroom lectures dealt with the variety of crimes the recruits would face as police officers, with stress on the proper method of filling out forms, one of which seemed to exist for every felony and misdemeanor man was capable of, and every mishap that might befall him. This emphasis on paperwork weighted things heavily against imaginative, action-oriented patrolmen. The men who kept abreast of departmental rules and procedures, which were always being revised and amended, usually ended up with clerical assignments in a precinct; such desk jobs gave them enormous power, simply in knowing how to prepare and route the stream of reports that flowed daily through every station house, and invariably they were the ones who did the best in the civil-service exams for promotion to sergeant, the first big step up in the police chain of command.

"In a vehicular collision between private auto and taxi in which the private auto operator is deceased," Serpico wrote dutifully in his notebook, "call desk officer for referral to precinct detectives, homicide squad, and accident investigation squad. Interrogate witnesses and make necessary entries in memo book, including list of deceased's property for desk officer. Fill out form UF 95 and attach to deceased's right big toe. Get receipt from morgue wagon operator. Fill out UF 6 for precinct file. Prepare UF 6 duplicates for Information Unit, Accident Record Bureau, and Hack Bureau. Fill out detailed accident report UF 6(B). Fill out MV 104 for State Motor Vehicle Dept. Do not fill out UF 18 unless there is city liability. If so, make copy for Corporation Counsel." Later, when Serpico was assigned to a precinct, he was advised by a veteran patrolman, "If you hear the sound of a crash on your post, the smartest thing you can do is run the other way."

Besides the firing range, where he received an "expert" rating, he most enjoyed fieldwork, actually observing cops on the job. But the high point came unexpectedly after he was three-quarters of the way through training. In an emergency, recruits can be placed on duty in the street, and when a rash of synagogue desecrations erupted throughout the city, Serpico's class was called out. The men operated in teams of two. They were permitted to choose their partners, and each team was assigned a synagogue to watch. The recruits wore civilian dress, worked only at night, and could use their own cars if they had them. Serpico had naturally gravitated to the former department-store detective, a wiry Irishman named John O'Connor, who shared many of his motivations about being a police officer. The two of them had often gone out for dinner together and then scouted around hoping to spot a burglary; now, each time they set out on the synagogue detail in Serpico's battered Lark sedan, O'Connor would rub his hands together and say, "Tonight, Frank, my boy, tonight we make our first collar," but nothing materialized.

Finally one evening they were parked diagonally across the street from a synagogue on Manhattan's Upper West Side. It was late, the street deserted. On their immediate left was a school yard with a chain-link fence. They were chatting quietly. O'Connor, an intensely religious man who was disturbed that Serpico only sporadically attended mass, was saying, "Frank, it's the only true church. You can't turn your back on it." Suddenly, out of the corner of his eye, Serpico spied two figures darting across the school yard toward the fence. In the glow of the street lamp, he saw something glitter in one of their hands. They had climbed the fence and were just dropping over it when Serpico said, "Come on!" and he and O'Connor scrambled out of the car. They were waiting, guns drawn, when two young black men dropped to the sidewalk, turned, and faced them, speechless and out of breath.

It's really happening, Serpico thought, and identified himself as a police officer. O'Connor ordered them to face back toward the fence. One of the two men was short, the other tall. At the academy they had been told that this type of physical relationship was common among criminals. "Looks like a Mutt-and-Jeff team," Serpico whispered. O'Connor reached forward and took the object that had been glittering in the hand of the short man. It was a chrome money changer filled with coins. The short man claimed he had found it in the school yard. Serpico, meanwhile, searched the tall one and discovered a straight razor in a jacket pocket.

"What's this for?" he asked.

"Man, I need that for protection around here. I got robbed last week, and there was no cops around. Cops is never around when you *need* them."

O'Connor decided to try some psychology. "Look, fellows, we're going to have to bring you in for questioning. If you did anything wrong, we'll find out. But if you cooperate, we'll go easy on you."

"No, sir, it's just like we say," the short man said. "We found that money. Can we get to keep it?"

He sounded so convincing that O'Connor replied hesitantly, "Well, if nobody claims it, I guess it's yours."

"Oh, cool, man."

Serpico did not especially cotton to the idea of

suspects riding unfettered in the car. "John," he said, "we better cuff them." He could see the indecision in his partner's face, the embarrassment of a recruit possibly making an illegal arrest.

"Gee, Frank, I don't know."

The two men they were holding at gunpoint remained frozen against the fence. "Remember what that lieutenant told us the other day?" Serpico said. "If a felony has been committed, and an officer has reasonable grounds to believe the person apprehended has committed it, it's sufficient grounds to take him into custody. Right?"

"Yeah, but we don't know if a felony has been committed."

"In this neighborhood?" Serpico said. "You got to be kidding. I'll take the responsibility."

As they entered the local precinct house, they removed the handcuffs from the prisoners and approached the desk sergeant with some anxiety. But after he heard their story, he said, "Boys, you might have a good arrest. Bring them up to the detective squad room."

In the squad room a fat, elderly man was gesturing wildly with his cap to a detective taking notes, and when he saw the two men come in with Serpico and O'Connor, he yelled, "That's them! That's the guys that held me up!" The fat man was a cab driver and, as it turned out, one of a number of cabbies whom the pair had robbed in recent weeks. The two men would hail a taxi, direct it to some deserted section of the city, and then the tall one would hold a razor to his victim's throat while his stubby cohort rifled the driver's pockets for bills and took his money changer. Serpico and O'Connor had nabbed the two men within twenty minutes of their last caper, before it had even been broadcast over the police radio.

Elated by their triumph, they celebrated over a beer at the first saloon they could find. The next day, Serpico, as the arresting officer, had to show up for the arraignment of the two prisoners, and while he was escorting them into court, the short one said to him, "Man, you lucky you had your guns out. The way you two was carrying on, we would've been long gone."

The news of what Serpico and O'Connor had done spread quickly through the academy. Several instructors told them they were a cinch to win the Mayor's Trophy, the highest honor given out at the graduating exercises. In the end, though, another recruit, who had arrested a man for impersonating a police officer, got the award. The rumor was that he was related to a deputy chief inspector. "See," a recruit told Serpico, "it's not what you do but who you know that counts."

Perhaps so, he thought, perhaps not; maybe the other arrest was more important. In any event, he was too excited by the imminence of being a full-fledged cop to be bothered by it. But the reality of his future was marked by the moment when he was fitted for his regulation blue uniform. Recruits had to buy them from a list of approved tailors, and were given a purchase allowance of a hundred twenty-five dollars. Afterward, they had to appear for inspection at the department's Equipment Bureau. Serpico was a little worried about his overcoat; he thought it was too baggy.

"Nah," the inspecting sergeant told him. "You want it nice and loose. You might get lucky."

"I don't get it."

"Who knows?" the sergeant said with a wink. "A liquor store could be stuck up on your post, or a supermarket. It's always a good idea to leave room for a couple of bundles."

The day before graduation, the police chaplain addressed Serpico's class and spoke of the "still, small voice" within them that had led to their decision to become police officers. It was a career that was more than a profession, the chaplain said, it was a "calling."

At the graduation ceremony itself, Serpico sat stiffly erect, in uniform in public for the first time, his father, mother, and sister in the audience, listening to the remarks of the Police Commissioner. Commissioners came and went, but their Police Academy orations remained virtually indistinguishable, as if their speechwriters one after another had pulled out the work of their predecessors.

Serpico heard that he had received the best and most intensive police training in the world, that he had studied the law, the science of police work, police tactics, and "criminal *modus operandi*," and that he was now fully prepared to engage in the war against crime, to put what he had learned into practice on the streets of the city. His distinctive uniform would identify him as a public servant; people who had never noticed him before would be observing him closely; what he did and said would take on a new significance; and his "somewhat limited" world would be suddenly expanded as the eyes and ears of the community focused on him.

His role as a peace officer, he was told, had great moral, social, and political implications. His courteous, dignified, and impartial behavior both on and off duty was critically important in generating respect for law and government, since the number of contacts which the public had with the police was far greater than those with all other governmental agencies combined. He would be expected, therefore, to discharge his obligations under the "philosophy upon which our country is founded"—equality of man under law, the worth of the individual, the essential dignity and integrity of every human being.

Serpico believed every word of it. Except for a passing reference that his personal conduct "must be above reproach," there was no mention of corruption, of the money that would be dangled in front of him, inside as well as outside the Police Department, of his sworn duty to report it wherever he found it. Nor had there been any in his recruit training. One of his instructors had said, "We teach you according to the book, but every precinct has its way of doing things. You'll learn in the field."

In the room at Brooklyn Jewish Hospital, the shades were drawn. Frank Serpico lay on his side, naked, a sheet covering him to his waist, the tubes inserted in his chest still in place. His left eye remained swollen shut; his right one, half opened, stared vacantly into space.

Occasionally he would cough, and wipe the bloody

mucus from his mouth with a pad of gauze. The entry wound of the bullet, about the size of a dime, had crusted over. His ear continued to drain, the blood in the fluid giving it an ugly brownish color.

Only his immediate family was allowed to see him, and the first night he was there all of them came—his father and mother, his two brothers and their wives, his sister and her husband. They simply stood there for a few minutes. His sister, Tina, recalled looking at him quietly for a while, at his horribly distended and discolored face, mottled black and blue, and as she watched he heaved a great sigh and a tear emerged from the closed eye and slid down his cheek. She ran out of the room into the corridor and began to weep uncontrollably. Later, at home, unable to sleep, she wrote an emotional poem to her brother, which began:

> *To you, right is right,*
> *Wrong is wrong.*
> *And there is no other way.*

Under sedation, with his regimen of powerful antibiotics already started, still being fed intravenously, in a hospital room kept perpetually dark, Serpico had no sense of time and place.

Some months before he had gone on his vacation to Hawaii, to the island of Maui, with one of his girl-friends, the stewardess, and he dreamed about it. They had driven to a beach on the tip of the island. There was a sort of commune, people living in huts in the woods along the beach, sharing whatever they had. He walked along the beach and saw pillboxes left over from the war, and the peace symbols that had been drawn on them. He and the girl lay on the warm sand, the sun beating down, and he watched the huge waves rolling in. Sometimes, just before they broke, he could see fish swimming through them. They looked, he thought, as if they were made of neon, big neon fish, and he laughed with delight at the idea. Some of the people on the beach were nude, and suddenly he too stripped off his trunks. It was the first time he had ever done anything like that. A man came by and offered them a coconut he had opened, and Serpico remembered the cool milk spilling down his body. That night the beach people built a fire, and he and the girl sat with them listening to the guitars and the singing and the roar of the surf, and he had never felt so free.

By his fourth day in the hospital, Serpico felt a little better, was more aware of his surroundings, conscious of the flowers filling his room. In the late afternoon a nurse brought him a letter.

He noticed that it had been addressed to him in care of Greenpoint Hospital and had been forwarded to Brooklyn Jewish. It was postmarked February 4, and must have been mailed a few hours after he was shot.

At first glance it appeared to be nothing more than a jolly get-well card. It said, "Recuperate Quickly!" Except that "Recuperate" had been crossed out with a pen and someone had scrawled instead, "Die." Underneath, in the same hand, was, "You Scumbag." It meant a condom, and it was a cop's word, the worst thing one cop could call another.

chapter 4

On March 6, 1960, the day after he was graduated from the academy, Frank Serpico entered a dingy, red-brick building in eastern Brooklyn. He was carrying a clothes bag and a satchel with his gun belt, nightstick, flashlight, and extra ammunition, and he was trying hard not to show his excitement. Bracketing the wooden, glass-paned front doors were green-colored lamps that carried the legend 81 PCT.

Just inside the doors there was a large room painted two-toned green, the lower third of it dark, the upper part several shades lighter. A huge desk ran almost the length of the wall on the right, and he went to it and looked up at the officer, a lieutenant, perched behind it. The lieutenant was engrossed in a paperbound book, *Civil Service Tutor for Promotion to Captain, N.Y.P.D.*

Serpico stood there, nervous and expectant. Nothing happened. Finally he said as smartly as he could, "Patrolman Serpico, sir, reporting for duty."

The lieutenant reluctantly raised his eyes from the book and glanced at him with marked disinterest. "Oh, yeah, right. Listen, go up those stairs and see the roll-call man. First door on the left."

During Serpico's last two weeks at the academy, he had been in a state of increasing suspense over where he would be sent. His instructors had spoken of some precincts that were "country clubs"—in the quieter residential sections of the city, particularly Staten Island and parts of Queens and the Bronx—and of others that were "action" precincts. "You'll learn more in an action precinct in three months," he had been told, "than you will in three years in one of those country clubs."

To his relief, the 81st—the "Eight-one," in cop talk—was an action precinct. It was a triangular, drab area in Brooklyn, for the most part run-down two- and three-family houses along with a number of small businesses, a kind of no-man's-land between the Bedford-Stuyvesant ghetto and the white, predominantly Italian, neighborhoods of Bushwick, and it had a high crime rate—robberies, rapes, burglaries, auto thefts, homicides.

Serpico was immediately assigned to a squad. This determined his duty hours. Members of each squad worked one week on a day tour from eight A.M. to four P.M., the next week from midnight to eight in the morning, and so on. Life in the station house centered in the so-called sitting room, which was in the rear overlooking a tiny yard. The yard was used mostly to tie up stray dogs until a van came for them from the Society for the Prevention of Cruelty to Animals; bodies found within the precinct boundaries were also temporarily stored there, in a sheet-metal box, awaiting the arrival of a morgue wagon.

There were perhaps a dozen wooden chairs in the spartan sitting room, two long tables, and a bench. In a corner were a shoe-shine machine and a full-length mirror so that patrolmen could check their appearance before going into the street. Lining another wall were glass cabinets filled with circulars for wanted criminals, the precinct's "target of the month"— usually a known gambler—and sheaves of paper describing lost property and missing persons. After putting on their uniforms in the showerless locker rooms upstairs, the patrolmen gathered in the sitting room a half hour before going on duty. Most of the time was spent jotting down the latest alarms off the police teletype. Usually nobody bothered with the description of people wanted for various crimes committed in the last few hours; they were too vague to be of much help. Each man, however, was required to list at least ten of the automobiles reported stolen— the alarm number, the car's make, color, and license number—and to be on the alert for them while he was on patrol.

When the roll-call man appeared, the officers fell in. The foot patrolmen were given their posts, which averaged five and a half city blocks, and the radio-car

men their sectors. One radio car was always designated the "gofer" car. It was responsible for bringing
sandwiches and beer to the station house's administrative and clerical personnel, and "flutes"—
Coca-Cola bottles filled with liquor supplied by bars
in the precinct—to the lieutenants and sergeants.
Then the big oak double doors at the end of the sitting room were swung open, and the shift marched
out past the front desk, sometimes stopping there for
an inspection by the lieutenant, and into the street.

Upon reporting to the roll-call man for duty, Serpico was given a map of the precinct, with its various
posts and sectors, and was advised that the precinct
veterans would "fill him in" on what else he had to
know. And as he walked away from the station house
on his first tour, an older cop fell in beside him and
asked what post he had. When Serpico told him, the
cop said, "Yeah, that's a good post, kid. You can eat at
the deli. They're good on the arm."

"You mean I don't have to pay?"

"Well, you know, just leave the guy a quarter."

Food at the cheapest possible price seemed to be
one of the precinct's major preoccupations. A patrolman on post was allowed an hour to eat. The time he
could do this was set by the roll-call man, so that his
post would be covered in his absence. Theoretically
he could not go outside the post limits while having
lunch or dinner, but this was universally ignored. Serpico soon learned that there were six eating spots in
the precinct that offered policemen meals either completely "on the arm"—free—or at token cost. Some
owners were simply cop buffs, but they, as well as

those who were not, expected special treatment in return—such as swift, unquestioned action in case someone on the premises was accused of causing trouble, or permitting their customers to double-park illegally on the street. An especially favored owner was given a Patrolmen's Benevolent Association card; only patrolmen were supposed to have one, and it practically guaranteed immunity to parking tickets when placed in a car windshield. It could also, as Serpico would wryly observe during his career, get the car's tires slashed.

The proprietor of a large café in the precinct was the recipient of a PBA card. Serpico went to his place several times before he decided to stop eating his meals for nothing. His decision was not based on ethical considerations, although he had felt more and more uncomfortable sitting at a table in uniform, knowing he wasn't going to pay, imagining people at other tables eyeing him and knowing it too. But what galled him was that when he and the other cops went into the café, the owner tried to palm off leftovers from the previous day or an item on the menu that wasn't moving. The longer he brooded about it, the more demeaning he found it, and anyway he liked to eat well, so one day after finishing his meal, Serpico asked the waitress for the bill.

"Oh, no, it's OK," she said.

Serpico picked up a menu, checked the prices and left the amount he owed on the table along with a tip. He was halfway down the block when the owner caught up to him with the money. "Hey," he said, "take this back. We don't charge for cops."

"How come?" Serpico replied.

"Aw, you know, the boys. I don't charge the boys. They all eat here."

"What's the matter, don't you have to pay for the food you serve?"

"Sure I do."

"Well," Serpico said, "I get paid a salary. I got money to eat whatever I want. I don't need your hand-outs."

"Listen, come in and eat anything you want. I'll tell you what. I'll charge you what it costs me. I just don't want to make a profit off you. Agreed?"

Serpico finally went along with the compromise, making up for the difference by leaving a larger tip, and the transformation whenever he entered the café was magical. "Frank," the owner would say, "try the roast beef today, rare just like you like it. Taste it, believe me, you'll see." And the waitresses began hovering over him. "You need anything, Frank? More coffee? What can I get you?"

It was a small thing, perhaps, but in retrospect Serpico would see it as another indication of the growing estrangement between the police and much of the public, a breakdown of respect—a feeling that too many cops were taking whatever they could, not caring what anyone thought, whether or not it denigrated them, so that even the café owner, ostensibly a booster of policemen, was in fact treating them with disdain.

This could also be seen in the evictions that periodically took place in the precinct. Patrolmen were assigned to accompany the city marshals serving eviction notices, officially to maintain peace, but the marshals would give them five dollars and expect them to execute the evictions, to smash in the door of an apartment if necessary and run the occupants out. Serpico remembered the first time he went with a marshal. He refused the five dollars the marshal had proffered, but the marshal was so used to cops doing his bidding that, when they found the door locked, he handed Serpico a sledgehammer and said, "OK, officer, knock it down."

"Knock it down yourself," Serpico said. "That's your job."

After that, whenever he was handed a notice for an eviction during roll call, other patrolmen, knowing how he felt, would come up to Serpico and say, "Hey, Frank, you want me to take that for you?" A crowd always gathered during these evictions, and he wondered what it did to a policeman's image to be seen doing the marshal's dirty work. But in the station-house locker rooms a line he frequently heard was, "The public, what does the fucking public know?" and he would think of things like this later, when the alienation between the police and the public heightened, when he heard people in the streets dehumanizing cops, calling them pigs.

As a rookie cop, Serpico was also introduced to the fine art of "cooping," or sleeping on duty, a time-honored police practice that in other cities goes under such names as "huddling" and "going down." A policeman on post was supposed to ring into the station house at a specified time every hour. These calls were recorded by a sergeant manning the switchboard, but around three A.M. he was relieved by a patrolman. A sergeant, with much more to lose, would not risk falsifying the record of these calls, but the patrolman substituting for him did so as a matter of course if a particular officer let him know that he was going to coop during the night. Just to be on the safe side, many patrolmen carried little alarm clocks to wake them up when it was time to ring in, and it was considered a great joke to set a cop's clock so that it went off in his pocket during roll call before his tour began.

Some patrolmen started cooping as soon as the sergeant on patrol made his first "see"—police slang for a visual inspection to make sure all officers in the precinct were properly on duty. This usually occurred about one o'clock in the morning, and did not happen again until six, when a patrolman was required to check all the "glass"—shop windows—in his post.

One miserable, sleeting night in March, on his first twelve-to-eight tour, Serpico was standing by his call box when the sergeant drove up and asked, "How you doing, kid?"

Serpico saluted and replied, "OK, sarge."

"Well, I guess that's it," the sergeant said, pulling away. Perhaps an hour later, the sergeant, on his way back to the station house, passed Serpico again, bent against the whipping wind and sleet, and asked, "How come you're still out?"

"Huh?" Serpico said.

"Christ, you must be freezing."

"Yeah, it is a bit chilly."

"So go in," the sergeant said. "I gave you your 'see'. "

"In where?"

"Well, there's that school in the next post."

"Oh, right," Serpico said.

Out of curiosity he went to the school, and saw an empty radio car parked directly in front of it. Radio-car men cooped in high style, bringing folding cots with them. Some radio-car men preferred to coop in their cars, motors and heaters running, and the roll-up side windows had red stickers warning the occupants that if they could see the sticker, they were in danger of being asphyxiated; when the weather was warm enough, of course, they all slept in their cars, and Serpico never ceased to be amazed at their uncanny ability to rise out of a sound sleep the moment their number was called over the radio, mumbling, "That's us."

As Serpico stood across the street from the school, wondering what to do, he recognized the post patrolman coming out of the front gate. "You going in?" the patrolman casually asked.

It was all so open that Serpico could hardly believe it. He waited until the other cop had phoned in from his call box and accompanied him back into the school. The cop had a key for the gate as well as for the front door, and he led Serpico down to the boiler room. It was packed with policemen—some on cots, others stretched out on benches with their heads on inflatable plastic pillows, still others on the floor propped against what seemed to be milk crates. Another officer, rubbing his hands from the cold, pushed past Serpico into the room and started complaining bitterly about the lack of space. "Fuck you," a voice in the gloom said. "Next time get here early."

Depending on where one's post was in the precinct, there were other cooping spots available—a movie-theater lobby, the basement of an old-men's home, a maintenance shack in a park. The shack had a balky oil stove that would occasionally spew forth clouds of smoke that sent the men stumbling and cursing to open the windows, but it had the advantage of being a few feet from a call box. Although Serpico cooped there once or twice, he did not find it especially conducive to rest. Alarm clocks rang through the night. A patrolman would leap up, dash to the box to report in, and then come back exclaiming loudly, "Jesus, it's cold as a bitch outside." Everyone tried to settle down again before another man rushed out and returned muttering, "I got a fucking job. A sick call. Can you imagine—at *this* hour? Who's got my stick? Where's my goddamn hat? Get up, will you, you're on my fucking hat, for Christ's sake."

Except for nights of arctic weather, when he would come in and stand around for a few minutes to get warm, Serpico finally quit cooping. Instead he bought himself thermal underwear and socks and insulated gloves. It was, he thought, a hell of a lot better than putting up with the fetid air, the constant interruptions, the cacophony of snores, the groans and belches and farts that were an integral part of sleeping on the job. Besides, there was always a chance that he might catch a burglar.

In one respect, the 81st Precinct fulfilled all of Serpico's expectations. It was, as advertised, an action precinct, and he soon experienced the extremes of frustration and satisfaction in being a police officer.

About two-thirty A.M. Serpico, walking his post, was just rounding a corner that brought him onto Fulton Street, a broad thoroughfare that is downtown Brooklyn's business center but which, by the time it reached the 81st, was lined by a ragtag collection of small shops. Suddenly he heard the crash of glass shattering. It sounded nearby, but this was before the installation of mercury-vapor street lamps, and in the dim light of the old-style lamps he could not see any broken windows.

He was sure that it was a burglary, and that the burglar was probably still in a shop somewhere down the block. He only had to worry about one side of the street, since a housing project took up all of the other side.

The first thing Serpico did was to take off his coat so that its brass buttons did not glitter in what little light there was. He drew his revolver and moved cautiously along the sidewalk. As far as he could tell, the street was deserted, and he heard nothing.

Suddenly a man came out of an appliance store carrying a television set and practically walked into him. For a split second they stared at each other. Serpico was so close to him that he could see the man's grotesquely scarred face. One knife scar ran from his mouth to his ear, pulling his whole face out of proportion. There were two other purplish slashes down the opposite cheek.

Serpico was the first to react. "Freeze," he said to the man with the television set. "Don't drop it. Don't try to run." He moved back slightly so that the man could not throw the set at him. Then he ordered him to place it very slowly on the sidewalk.

"Lean against the wall," Serpico said. At the academy he had been told that most suspects, in order to be ready to make an attempt to escape, simply turned and faced the side of the building with their hands lightly against it. Serpico saw that the man was doing exactly that, and he could almost feel him tensing his body for the right moment to bolt. So he came up behind the man and snapped, "I said *lean*!" Revolver in hand, he hooked a foot around the man's ankle and pulled it back sharply. At the same time he shoved hard with his free hand against the small of the man's back.

The man grunted with surprise and said, "Watch out. I'll fall."

"You got the idea," Serpico said. "And remember, this gun's cocked."

With the man stretched out against the wall, Serpico frisked him and found a wicked-looking knife in his pocket; when he opened it, the blade was about six inches long. "You like to carve people up, huh?" Serpico said. The man did not answer. Then Serpico walked him to a call box on the corner and phoned in for a radio car. When the radio car arrived, his prisoner was handcuffed and put in the backseat along with the television set, and one of the radio-car men got out to guard the shattered storefront. An old man in a robe was peering anxiously out of the door. He identified himself as the proprietor. He and his wife lived above the store, and Serpico took his name and telephone number.

At the station house Serpico took the man into the back room and began filling out an arrest card. This had to include the prisoner's name, address, date and place of birth, and occupation if any, a description of the circumstances of the arrest, and the charges against him—breaking and entering, which was the burglary; grand larceny, which was the theft of the television set; and possession of a dangerous weapon, the knife. Then he took the man before the desk sergeant, where a brief entry was made in the precinct blotter that noted why Serpico was off his post, and one in the arrest record book. Since the 81st Precinct did not have overnight detention facilities, the switchboard telephoned for a van to take the prisoner to the adjoining 83rd Precinct, which did. In the meantime Serpico escorted him to the detective squad room for fingerprinting. At least once each night, and sometimes twice, the fingerprint cards were collected from precincts throughout the city and taken to the Bureau of Criminal Identification at Police Headquarters. This would determine whether the man had a record.

Because Serpico would have to accompany his prisoner to court in the morning, he was officially dismissed at five A.M. and retired to one of the bunks upstairs in the station house. These bunks were supposedly maintained by widows of slain policemen. The women were paid out of the monthly house-tax contributions—a dollar eighty-five when Serpico was in the precinct—that went for coffee and rolls, the police magazine, and the so-called widows' fund. The only person he ever saw making up the bunks, however, was the "broom man," an elderly cop who served as the station-house janitor and sometimes cooked hot meals for the clerical staff on a stove in the basement.

Shortly after eight A.M. Serpico arrived at the 83rd Precinct to collect his prisoner. He climbed into a wagon with other officers and their prisoners, and drove to the photo unit at the Brooklyn Borough Command for mug shots. There he was given a sheet sent over from the Bureau of Criminal Identification listing his prisoner's previous arrest record. The cop who handed the sheet to Serpico said, "You got a good one."

The man had served five years for sticking up a gas station, two years for assault and robbery with a knife, and a year and a day for simple assault with a knife. This was his fourth felonious offense, and as a "three-time loser," he faced twenty years to life in jail. "You must be slipping," Serpico told him, "going for a TV set."

"You got to live," the man said. "You got to do something."

Serpico took him to the men's House of Detention, next to the courthouse, to wait for his arraignment. It was not until three in the afternoon that the man, unable to make bail, was remanded to the detention house for trial, and Serpico's day was done. About three weeks later he appeared in court again for a pretrial hearing. By now the man had a lawyer. Serpico watched while the lawyer, the judge, and the assistant district attorney prosecuting the case conferred. Then, in disbelief, he heard the charges against the man reduced to a single misdemeanor, unlawful entry. The man stepped forward, pleaded guilty, and the judge sentenced him to three months.

Serpico went up to the assistant district attorney. "What's all this bullshit about a three-time loser?" he said. "The guy's a bad guy. He's committed three felonies."

"You don't want him to do twenty years for a lousy television set, do you?"

"Listen, you saw the knife he had. There was an old man and his wife living there. What about what he might have done to them?"

"Yeah," the assistant district attorney said, "but he didn't."

Serpico began to share—as well as understand—the disillusionment about law enforcement that was behind the cynicism of so many police officers. But a few days later an incident occurred that alleviated his anger and made him feel very good again about being a cop, made him feel there was nothing he could do that was more worthwhile. He had just finished handling traffic at a school crossing when a young black man ran up to him and said breathlessly, "Officer, come quick."

"What's the problem?"

"Come quick, *please*. My wife, my wife's having a baby."

Serpico followed him around the corner and up the stairs of a tenement into a sparsely furnished two-room apartment. "It's our first," the man said, "and it's happening so fast I don't know what to do."

Serpico looked in the bedroom. The young man's wife was in bed under a blanket, her wide-eyed, frightened face staring back at him. He guessed that she was not more than twenty, and from her groans and the contractions of her body he could see that she was in deep labor.

He took off his coat and went into the bathroom to wash his hands. He asked the man if he had a phone. The man did not, and Serpico said quickly, "Well, find one and call an ambulance." Then he went back into the bedroom. "It's OK," he reassured the woman, "you're going to be OK." She uttered a piercing little cry of pain and began biting her knuckle. Serpico pulled the blanket aside and saw that she was not only in labor, but that birth was imminent. She had already broken water.

There had been instruction at the academy about emergency deliveries, along with a training film, and Serpico ran over in his mind what he had been taught. To his surprise, he was not nervous at all; he remembered, instead, feeling a sense of elation that he was on hand to aid this woman lying helpless and afraid on the bed, dependent upon him, trusting him, that he was going to deliver her baby, see it being born, starting life.

He saw the baby's head beginning to emerge, and he cupped his hand under the head to support it. The birth was going to be easy, he thought, no problems. Then, suddenly, he saw that the baby was on the verge of strangling. The umbilical cord was wrapped around its neck, and as more of the baby came out, the tighter the cord got. With one hand holding back the baby's progress as much as he could, Serpico worked frantically to slip the cord over its head. Finally he succeeded.

After the baby had been delivered, he turned it upside down, cleaned out its mouth with his finger, and gave it a slap to start it crying. "What do you know," he said to the woman, "it's a boy."

He had been told at the academy not to cut the umbilical cord unless absolutely necessary, to let the ambulance personnel do it, and he wondered where the ambulance was. Still holding the baby, he turned and saw the new father clutching the doorway, transfixed. "Have you been here all the time?" Serpico said.

The young man nodded dumbly.

"You didn't call the ambulance?"

He swallowed hard. "No, sir, I was, you know, too scared to leave. I won't even have to pay the doctor now."

Serpico told him to boil some string and a pair of scissors. He placed the baby on the mother's stomach, and using the span of his fingers as a guide, he tied a piece of the sterilized string around the cord approximately six inches from the baby's navel, and another one two inches farther, and cut the cord between them. Then he wiped off the baby with a towel and put him in his mother's arms.

Serpico attended four more births as a patrolman, but this one, the first, was the one he would always remember. About a week later, he was walking a post when the baby's father appeared. "Officer! I been looking all over for you."

"What's the trouble now?"

"No trouble. Baby's fine. Everything's fine. I just want to know your name."

"Serpico."

"No, I mean, like, your whole name."

"Frank. Frank Serpico."

"Well, that's what we're calling him—*Frank*. You know, after you."

Except on really foul nights in the middle of winter, Serpico loved walking a post, the feeling of freedom he had when he turned out of the station house into the street, the sense that he was on his own, in control, responsible for the safety and well-being of everyone in the five and a half blocks of the city he would cover during his tour of duty. He was essentially outgoing anyway, and he took very seriously the precept set forth at his graduation from the Police Academy that he was the personification of government for most people, so he always smiled and nodded at passersby, stopped to chat with shopkeepers on his post, and made a point of helping elderly ladies across the street.

But as far as law enforcement was concerned, his job as a uniformed cop was basically limited to preventing crime or apprehending criminals on the spot; anything beyond this, the investigative aspect of police work—sifting for clues, following up leads, breaking down a suspect, cracking a case—was handled by detectives, and he began to dream more and more of the day when he would be one himself. About all he could practice in the way of detection while walking a post was to search for stolen cars, and he became the top man in the precinct in recovering them. What Serpico did was to concentrate on license plates. Every time he looked at one, he ran through a checklist of suspicious signs in his mind. Was a plate dirty and the car clean? Did the plate screws have nicks and indentations in their heads? Was there rust around an empty screw hole, indicating that a screw had once been in it and then removed? Were the screws loosely in place, or the license plate tied on with wire? Each was sufficient grounds for questioning the car's driver. Sometimes an owner had simply been careless when transferring his plates, but more often than not either the car or plates turned out to have been stolen.

This constant determination to excel as a police officer, to fulfill his boyhood dream of what a cop should be, finally enabled Serpico to play detective in a case very different from a car theft. Along with other crimes in the 81st Precinct, the incidence of rape was high, but most of the men stationed there had a calloused attitude about it, particularly when it involved black women. "They're always hollering 'rape,' and half the time they're asking for it," a veteran detective told him. One day Serpico was on a post when he heard a muffled cry in a cellar. He hurried down the stairs and saw two black figures, male and female, grappling on the floor. The man jumped up and ran out of a door leading to a courtyard. Serpico could have taken a shot at him, but he remembered what the detective had said and did not. As it happened, the woman not only had been drinking, but refused to make a complaint. "See," the detective reminded Serpico, "you got no complaint, you got no case. You fire your gun, you got to go down to ballistics, and that's a pain in the ass. You hit some bystander by mistake, then it's your ass all the way. And what for? To get some fucking guy who's supposed to be raping some nigger bitch?"

One night, however, when Serpico was filling in for a radio-car man on sick leave, a call came in at about eleven P.M. that a woman was screaming in the street along the borderline of the sector he and his partner were covering. As they were responding to the call, the car in the next sector arrived at the scene and reported, "Condition unfounded." Rather than turn around, Serpico, who was at the wheel, continued cruising toward the point where the screams had been heard. There was a school yard on the left. "I've been on post here a lot," Serpico said. "That yard runs back pretty far. Let's take a look." He swung the car up on the sidewalk so that its headlights swept the school yard, and in the dark rear of the yard the lights picked up four youths surrounding someone on the ground. They immediately started scrambling away in different directions. Serpico and his partner ran to the form on the ground and saw that it was a black woman, moaning incoherently, stripped naked. Then they went after the youths. Three of them escaped through holes in the yard fence. The fourth was trying to climb the fence, had just reached the top of it, when Serpico hurled himself at him and grabbed one of his legs, and pulled him back into the yard and handcuffed him.

The dazed woman was helped to her feet and wrapped in a blanket from the radio car, and Serpico retrieved her torn panties as evidence. She numbly explained that she had been sitting in a car alone while her boyfriend delivered a package and had suddenly been dragged out into the school yard and assaulted. Many rape victims, whether black or white, did not press charges because of the social stigma, the fear of retribution, or the humiliation they often had to endure, especially if they were young, in station houses—the appraising glances, the whispered comments. But when the woman was brought in to identify the eighteen-year-old youth Serpico had captured, she was overcome with rage. "That's one of them!" she shrieked. "He was holding me when this other one put his thing right up to my mouth and said if I didn't, you know, do it, he'd kill me." Oh boy, Serpico thought, she's going to make a fantastic witness.

The woman was taken to the hospital, and Serpico had to turn the prisoner over to the precinct detectives for questioning about his confederates. There were three detectives in the squad room, one of whom was nicknamed "Blackjack" because of his enthusiastic use of that leather-bound lead instrument. "OK, you fuck, who were your playmates?" the detective called Blackjack began. When there was no answer, he took his blackjack out and held it in his fist so that about two inches of its business end protruded, and rammed it into the rapist's stomach. He crumpled over with a soft sigh. The blackjack came whistling in again and again, to his stomach, his kidneys, the back of his neck. He passed out. A pitcher of water revived him. He was put in a chair, and one of the other detectives came up behind him with a telephone book and slammed it against his ear. As the youth slowly shook his head in pain, the telephone book pounded his other ear. Back and forth it went. Occasionally the telephone book missed an ear, and his nose began spurting blood. Still he would not talk, and finally Serpico was told, "Get him the fuck out of here."

As the arresting officer, Serpico picked up his prisoner in the morning to escort him to the photo unit for mug shots and then to court. But instead of getting in the van with him, Serpico followed in his own car. He thought the detectives the night before were much more interested in beating the prisoner than in getting information out of him, and he had an idea. So after the photographs were snapped, he broke a departmental rule. A prisoner was supposed to be put in one of the wagons outside to go to the Men's House of Detention to await arraignment later in the day. But Serpico took him aside and asked, "How you feeling?"

The boy, handcuffed, cowered as if he expected a kick in the groin. "No, I really mean it," Serpico said. "They worked you over pretty good last night. There wasn't much I could do about it. Say, are you hungry?"

He remained silent, suspicious.

"Look," Serpico said, "there's a cafeteria across the street. I'm going to take you over there with the cuffs off. I want you to get one thing straight, though. I didn't lay a hand on you last night except when I grabbed you, but you try to split on me now and I'll put one right in your back."

As they went into the cafeteria, Serpico gave him two dollars. They sat at the counter, and the boy ordered eggs and coffee, but he couldn't get the eggs down. Then he spoke to Serpico for the first time. "Can I have some cigarettes?"

"It's your two bucks. Spend it any way you want."

When they left, Serpico decided to risk driving the boy to the detention house instead of putting him in a wagon. But he did not want him in the car without handcuffs, so he told him what he was going to do and said, "I have to put the cuffs back on. Anybody spots us, I can get myself in a lot of trouble."

Serpico tuned in some music on the radio, and after a few moments said, "I got to hand it to you. Man, you can really take it. You must be hurting all over." He paused. "It's too bad you have to take the whole rap for those other guys. Like, I don't know if you were the most guilty or not, maybe you just went along, but you're the one who got caught."

The boy did not respond. Out of the corner of his eye, Serpico saw that he was looking straight ahead, puffing furiously on his cigarette.

Serpico pulled his car up in front of the big steel gates of the detention house. "Well," he said, "I guess you better take your last free breath because it's going to be a long, long time before you take another. I don't understand it. You feel obligated to those guys or something? Fuck it, they ran out on you, left you holding the bag. They're probably laughing right now: 'Old dildo got caught, but fuck *him*.' You know it's really going to be bad on you. You know what you can get for this?"

His prisoner shifted nervously in the seat, took another deep drag on his cigarette, and mumbled, "OK," and named his three accomplices.

"I'll talk to the D.A.," Serpico said. "I'll tell him you cooperated."

That evening, although he was off duty, Serpico went back to the school, which stayed open after classes so that its gym and yard could be used as a neighborhood youth center. Serpico knew the man in charge, and showed him the names. "Sure, I know them," he said. "They're around here all the time."

Serpico called the detective squad room, explained where he was, and said, "Send someone over. I got a line on those guys in the rape case last night. They'll be here any second."

The detective replied, "Gee, Blackjack's catching that case, and he's off for a couple of days."

"So what?"

"Well, what can I do? It's not my case, you know."

The hell with them, Serpico thought. While tracking down suspects was the province of detectives, he had gone this far and he might as well go all the way; and the following day, back on foot patrol on the four-to-midnight tour, he asked the roll-call man to assign him to the post that included the school. He also told the officer on the adjoining post to be near the school at seven o'clock. "I think I got some collars coming up," he said, "and I may need you."

He went to the school and waited until two of the suspects showed up. One of them sat down on a bench in the yard to watch a softball game. The second strolled into the gym. Serpico sauntered around the yard until he was behind the one on the bench. He could hear the youth's breathing quicken and he said, "Sweat, you son of a bitch. You move an inch and

I'll fucking well kill you." The cop from the next post was standing outside the fence, and Serpico signaled him to come in. They took the suspect inside, and then Serpico went into the gym for the second one. He found him lounging against a wall, ogling a couple of girls. "Come on," Serpico said, "I want to talk to you."

"I didn't do nothing. I didn't do nothing. What for you want me?"

"Three guesses."

Serpico telephoned the woman who had been attacked and asked her to meet him at the station house. When she arrived there were no detectives in the squad room so he conducted his own lineup. The woman looked through a one-way glass and made a positive identification of the prisoners.

By then a couple of detectives had wandered in, and Serpico gave them the name and probable address of the fourth suspect, who had not shown up. He took the woman downstairs, filled out more forms relating to the case, and had a cup of coffee.

Possibly three-quarters of an hour passed before he returned to the squad room. One of the detectives told him that the fourth man had not been at the address, but that through an informant's tip he had been nabbed on the street and would be brought in shortly. Serpico straddled a chair to await his arrival. Almost at once he sensed that something was wrong; the squad room had become very quiet. The detective who had spoken to him before busied himself with some papers. Then he eyed Serpico for a moment and

suggested that there was no point in his hanging
around any longer.

"What do you mean?" Serpico asked him, puzzled.

"Kid, it just can't be," the detective said.

"What can't be?"

"We got to take the collar, the whole thing. We got
the squeal, and it's our case, and it wouldn't look
right—you get my meaning?—if you came up with
everything."

"Fuck you," Serpico said. "I broke my ass on this
one. I even called the office last night to tell you where
to find these guys, but you couldn't be bothered."

The detective tried to reason with him. He reminded Serpico of how furious one of the lieutenants
had been the week before when a patrolman had arrested a man the detectives were looking for. "Jesus,"
he said, "the boss almost had a hemorrhage."

"This is my collar, and I'm taking it."

The detective sighed and said he would have to
phone the lieutenant. Serpico listened as he sketched
out what had happened, and heard him stammer,
"Hey, boss, don't get sore at me. It's not my fault." The
detective hung up, and said, "Well, he's real pissed off,
and he's coming down. So stick around."

When the lieutenant came in, he ordered Serpico
into his office, informed him that he was conducting
an official inquiry, and, with pen in hand, started out
by asking him his name, rank, and shield number.
Then he demanded to see Serpico's memo book. A
memo book is in effect a patrolman's diary of what has
occurred on his post. Theoretically it is supposed to
be kept up to the minute, and that was the one thing

that Serpico, in all the confusion, had not yet filled in. "Hah," the lieutenant said, scribbling it down, "no memo-book entry." Next he wanted to know why Serpico was off post, why he had left the street to go into the school, without permission. Technically this was another violation of the department's rules and procedures.

"What are you trying to tell me, lieutenant, that you want the collar?" Serpico finally said.

The lieutenant replied that it was either that or a "complaint"—a charge that Serpico had broken regulations. The upshot would be a departmental trial, the equivalent of a court-martial in the army. Serpico, as a rookie cop, was so intimidated by the prospect of defending himself against a lieutenant, and so aghast at the idea of his record, at the very least, being marred by a reprimand, that he gave up.

"OK," he said, "you want the collar, you got it," and walked out.

There was a ludicrous windup to the incident. Some of the detectives tried to jolly Serpico out of his anger the next day, but by then he had told a number of other patrolmen what had happened, and word of it had reached the precinct captain. He questioned Serpico, and stormed up to see the lieutenant. He returned to tell him that he was sorry, all the reports had been filed and it would be too embarrassing for everybody to change them. But, he said, they had agreed to give Serpico an "assist" on the arrests, and added that he would be "written up" for a commendation.

Serpico could hardly recognize himself in the report that earned him his first commendation for

Excellent Police Duty. To keep everything consistent with what had already been recorded, it made no mention of his investigative work that had led to the three arrests after the night of the rape. It stated, instead, that Serpico and his partner in the radio car had made the initial arrest at great personal risk since "one or more perpetrators" of the crime were believed to have been armed with knives.

Excellent Police Duty is the department's lowest award, and if in his appearance before the Honor Board Serpico had not said that he was unaware of any knives flashing in the school yard, he probably would have received a higher one. What he ended up with was bestowed on the grounds that the two officers had saved the victim from more serious bodily harm.

Serpico's education as a police officer in the 81st Precinct also included his first brush with graft. It happened when he was substituting again for one of the regular radio-car men.

Serpico was particularly sought after as a replacement because he was always willing to handle the wheel during the entire eight-hour tour, instead of the usual procedure of dividing up the driving time. Actually it was no hardship for him. He preferred to drive; it gave him a sense of being on top of things, of being his own man. And so one day, when he spotted a car ahead of him going through a red light, he simply sped up without a word and pulled the car over. While his partner, a big-bellied veteran, lolled back in his seat, Serpico got out and asked the driver for his license and registration. The license had several violations listed on it, and the man was beside himself. He told Serpico that he had not seen the light change until he was halfway through the intersection. He was a salesman dependent upon the car, and if he received another ticket he would lose his license altogether. "Can't you give me a break?" he pleaded.

The car had not raced recklessly past the light, had rather eased through it, as if the man, as he said, had not been paying attention to what he was doing; and Serpico was on the verge of letting him go with a warning when the man said, "Look, it's worth thirty-five dollars to me not to get a ticket."

Serpico stared at him. "You want to give me thirty-five dollars not to give you a ticket?"

"Officer, that's all I've got in my wallet. Honest."

"Wait here just a minute," Serpico said. He went back to the radio car and said to his partner, "Hey, come out and hear this. I need a witness. I'm going to lock up this guy for bribery."

"Bribery? What happened?"

"The guy says he wants to give me thirty-five dollars if we let him go for this light."

Serpico's partner said, "Uh, get in the car. I better handle this one."

As the junior man on the team, Serpico had to defer. He waited in the radio car for a minute or so, then watched his partner return while the man drove off.

"What's going on?" Serpico asked.

His partner got back in the car. He opened his hand and counted out three crumpled tens and a five. "You got change? You know, we split fifty-fifty."

Serpico started up the car. "No," he said blankly. "I'm really pretty independent. I don't need the money."

"You sure? Real sure? I mean, you know, you're supposed to get half. We split everything right down the middle."

"It's OK," Serpico said. "You let him go. You made the score. You can keep it."

"Boy, Frank, that's nice of you. This'll buy a lot of milk for the kids."

The episode, when it got around, made Serpico even more sought after as a partner among old-time radio-car men. And when he got over his initial shock he tried to rationalize what had happened. After all, as a kid, he had seen cops helping themselves to fruit from the neighborhood markets, and everyone took this as a matter of course. The traffic incident was just an extension of the same syndrome. Whether it was an apple or a traffic-ticket bribe, some cops had simply come to expect it as their natural due, and probably didn't give it a second thought. Besides, it was the man in the car—the public, one might say—who had started the whole thing; he was the true corrupter. In any event, what counted most, Serpico told himself, was that *he* had not taken the money, the other cop had, and thus it was none of his concern.

Even if he had thought differently, what could he do as a rookie cop in the precinct? He could, of course, go to the desk sergeant or lieutenant and report that his partner, a member of the force for more than ten years, had accepted a bribe. But for all he knew, they were aware of it, or possibly taking money themselves on occasion. There was no real machinery to deal with the problem, nor any encouragement to do something about it. Nobody sat around talking about corruption. It was all very amorphous, just there, permeating the department. Some cops were on the take, others were not. It was a personal decision every rookie had to make sooner or later. The choice, however, was limited to participating or looking the other way. Anything else was unthinkable. It would break the code.

Serpico did mention the incident to one person, however, a girl he was dating at the time, a receptionist for a Manhattan advertising agency. He had met her when she had a similar job with an employment agency, where he had gone job-hunting while waiting for his appointment to the Police Academy to come through. They had started talking, he had told her that he really wanted to be a cop, and it turned out that she had an uncle who was a patrolman and a cousin who was a detective. After they began going together, they bickered constantly. She drove him nearly crazy with talk about her glamorous life at the ad agency, about how any day she would become a model, that he spent too much time playing cop off duty and did not pay enough attention to her, that he ought to get a new car, that he never bought her a decent present. He swore again and again that he would break off with her, but her body was as lush as her mind was mean. In bed she became another creature, possessed, demonic, and it was then that his resolve to leave her would evaporate.

The night he told her about the bribe, she lay on the bed, holding a cigarette between her lips, and said, "Why didn't you take the money? Everybody else does."

Still, he believed then—and would believe until the very end—that there was a mystique that did bind cops together, that there were moments of absolute unity when it did not matter whether a cop was crooked or honest. And these moments happened when another cop was in trouble. He would never forget one summer night around two A.M. He was on post in the 81st Precinct and heard what sounded like the crash of metal behind him. As he started back toward it there was an odd thumping noise, and then he saw this big white Cadillac creeping the wrong way up a one-way street at about five miles an hour. He fixed his flashlight on the car, and saw that one of the front fenders was badly dented. "Hold it!" he hollered. The car stopped, and he approached it on the driver's side. There was a barrage of shouts and curses from inside the car demanding to know why the driver had stopped, telling him to get going again. Serpico moved more cautiously, his hand now on his revolver. When he got to the driver's door, he opened it slightly so that the interior lights went on. The light brought on another round of shouts and jeers. He looked in and saw five very drunk men. One of the men in the rear seat started groping for the door handle, trying to get out. "Police," Serpico said. "Everybody stays in the car." This triggered more jeers and protests. "OK," he ordered, "keep it down." The noise in the car subsided to angry, confused muttering, and Serpico told the driver to give him his license and registration.

"What for?" the driver said, his speech slurred. "I didn't do nothing."

"Yeah, I can see that from your fender. Come on, let's have them."

As the driver fumbled through his wallet, Serpico heard the whine of sirens in the distance. It got louder, then ear-splitting. Suddenly the street was filled with radio cars, tires screeching, lights flashing, the whole street ablaze with lights, people hanging out windows, and cops—it seemed to him like dozens of them— running toward him, guns drawn.

It was unreal, as if he were watching a movie, and for a second he was as dazed and surprised as the frightened occupants of the car.

"You all right, kid?" a sergeant said.

Later Serpico learned that someone on the block, hearing the commotion after he had stopped the car, had telephoned the police, and a signal "ten-thirteen" had gone out over the radio. It meant "assist patrol-man," that he was in trouble. Serpico explained that he simply had a drunk driver and a possible accident. His surprise gave way to embarrassment, and he began to apologize for the unnecessary bother he had caused.

"Don't feel bad about it," the sergeant said. "You never know when it's the real thing."

Afterward he reflected on how he had been alone on the street, and how they came at once, unques-tioningly, thinking that he needed them, and the memory of that moment remained with him always, like the lost innocence of a child.

It was in the 81st, too, that Serpico first used his gun, deliberately took aim at another human being and pulled the trigger. Around eleven o'clock in the evening he was walking his post and stopped to chat with another officer on an adjoining beat. As they talked, a man ran up to them and said that a house was being robbed down the block and across the street. "Let's go," Serpico said.

"Wait a minute," the second cop interjected. "That side of the street isn't in our precinct."

Serpico had already encountered this sort of bureaucratic thinking—it seemed to permeate the department, the idea of doing as little as possible, of playing it safe. "Are you kidding?" he snapped. "That shield you got doesn't say the 'Eight-one Precinct'; it says 'City of New York.' Come on."

The second cop reluctantly followed. The street dividing the two precincts was residential, mostly one- and two-family houses with fenced-in backyards. The house in question was set back slightly from the sidewalk. Serpico entered an iron gate and went up to a ground-floor window. The room was dark, but a door to another, lighted room was open, and he could see a man in it rummaging through a chest of drawers. Then he noticed a second, raised window, partially hidden by some shrubbery.

He came back to the sidewalk and said to the man who had reported the robbery, "Who lives here? You?"

"No," the man said excitedly. "I'm only a neighbor,

but I know these people and I know they're away. That's why I got suspicious when I walked by and saw a light."

"OK," Serpico said, "you're to be commended. Now take it easy and stay out of the way." He told the second cop to guard the front of the house and he returned to the opened window, climbed through it, and tiptoed across the room toward the room where his quarry was. He must have made some inadvertent noise, however, because just as he was edging through the doorway, the lights went out. Serpico's career as a cop might have ended right there. But some instinctive reflex action caused him to hunch over so that his left shoulder, rather than his head, took the brunt of what turned out to be a wrought-iron candelabrum that came crashing down on him.

The blow sent Serpico reeling against the wall, but he managed to recover, and even grappled briefly with his assailant before the man broke free and ran up the stairs. Serpico staggered after him, yanking his gun out of his holster. In a rear room on the second floor a window was open. As he peered out of it, he saw the figure of a man straddling a fence, his body silhouetted against a street lamp shining down an alley diagonally across from the window. "Police," Serpico shouted. "Stop, or I'll shoot."

The man kept going over another fence. Serpico sighted and fired. He thought he heard a muffled cry, but nothing more, no sound, no movement, and rather than scaling the fences himself, he decided it would be quicker to go around the block to the alley. On his way out, he met the second cop, still in front of the house, not knowing quite what to do. "Jeez," he asked, "what happened?"

"Come on," Serpico said. "Maybe we can catch him."

When the two officers reached the alley, they moved cautiously along it, revolvers in hand. But both the alley and the yard it led to were empty. Then, near the fence, Serpico's flashlight picked out traces of fresh blood and bits of flesh. "It looks like you hit him," the second cop said.

"Yeah," Serpico replied, "but he got away."

Since a shooting was involved, Serpico had to report it, and had to wait until a photo unit arrived to take pictures in the house to show that a burglary had been in progress. After all this, somewhat dejected, he went back to his precinct house. Nothing there helped his mood. Word of the episode had preceded him, and as he was signing his return roll-call sheet, a cop said derisively, "Hey, Serpico, I thought you were supposed to be such a great shot."

Worse was yet to come. The desk sergeant called him over. "I ought to put in a complaint against you, Serpico," he said. "All the time you're screwing around in another precinct trying to play hero, a guy on your own post got robbed and assaulted. They damn near cut his balls off."

"*What?*" Serpico said.

"Take a look for yourself. The guy's laying upstairs right now waiting for the ambulance."

He dashed up to a squad room on the floor above. A man, with some detectives watching him, was slumped in a chair, moaning, his hands clutching his bloody testicles.

Serpico was aghast. Finally he found words. "That's him! That's the guy I shot!"

"No, no," the man said. "They cut me."

"Are you sure?" one of the detectives said to Serpico.

"Of course I'm sure."

"OK, wise guy," the detective said, "we'll pour a little salt down there and see what you have to say then."

Sickened by the scene, Serpico left and went to the locker room. He was sweating profusely, his bruised shoulder was throbbing, and his head was aching badly. He sat there alone on a bench for perhaps twenty minutes before he pulled himself together and went out to the front desk, wondering if he would have to accompany the man to the hospital.

As he stood there, another detective approached him and said, "You can go home. We'll handle this. We called the Seven-nine Precinct and they came over. They had a warrant out for this guy."

"Yeah?"

"Yeah, and he won't be committing his specialty any more. You really hit him where it hurts. He's wanted for rape."

chapter 5

The second "condolence" card arrived at Brooklyn Jewish Hospital on February 10, 1971, seven days after Frank Serpico had been shot in the head. It was the kind that someone, lacking words of his own, might send to a bereaved friend or acquaintance. The printed message read:

WITH
SINCERE
SYMPATHY

Except this particular card contained an additional handwritten note of regret *"That you didn't get your brains blown out, you rat bastard. Happy relapse."* The drainage from his ear continued, but it had slowed somewhat. The heavy doses of antibiotics had thus far warded off any infection, and there was every hope that the rip in his cerebral membrane was beginning to mend. If it did not, the alternative was surgery, a possibility that had not yet been ruled out. The operation was a delicate one in which the neurosurgeon would have to go into the side of his head at the base of the brain, and, like any surgery of this sort, it carried with it great risk, especially unpredictable and uncontrollable swelling of the brain.

The left side of Serpico's face remained paralyzed, and his mouth was still twisted, but he could open the right corner of it enough to communicate fairly well and to down the diet of baby food he was on, now that he was no longer being fed intravenously. The worst part was the blinding headaches that came on him without warning.

He had asked for a mirror, and saw the clotted bullet hole on the left side of his nose. It seemed to him to be absurdly innocuous compared to the rest of his battered face. He looked as if he had been in a really good fight. While the swelling was nearly gone, there was still considerable discoloration. His left eye was blackened and still partly closed. With his beard, the overall effect, he thought, gave him a piratical air. One of his nurses tried to talk him into getting rid of the beard, but he refused; he rather liked the buccaneer image he presented.

The afternoon that he received the second anonymous card he was wheeled out of his room on a stretcher for a new series of X rays called a tomogram. This involved a lengthy process in which consecutive exposures were taken almost centimeter by centimeter around his head so that the neurosurgeons would have a more accurate, three-dimensional picture of the bullet fragments inside. It confirmed that the fragments were all in the area of the ear bone, and that one of them was very near the carotid artery. This fragment, and the possibility that it might move, caused the gravest concern. But the tomogram showed that so far it appeared to have stayed in its original position.

Serpico was under the direct care of the chief of neurosurgery at Brooklyn Jewish, Dr. Aaron J. Berman, and the associate chief, Dr. Zeki Ugar, a former medical officer in the Turkish navy who had been sent to the United States to study, and who had decided to return here after his naval service was over. Dr. Ugar had never met an American policeman, much less one who looked like Serpico, and, fascinated by his patient, had taken an interest in him beyond his medical condition, making it a point to drop in every afternoon to chat with him for a few minutes in an effort to boost his morale.

He came in after the tomogram had been taken to tell Serpico how encouraging the results were. But Serpico had something else on his mind. During most of his first week in the hospital, he had lain on his right side, oblivious to his surroundings. For the last two days, however, he had been more aware of what was going on, and his bed had been raised for brief periods. He noticed that whenever someone spoke to him, he had to turn the right side of his head toward the speaker to hear anything. He thought that it was caused perhaps by the blood clogging his left ear, and he now asked Dr. Ugar how soon it would be before he started getting his hearing back.

The problem had already been noted in Serpico's records, and Dr. Ugar knew that it was a post-traumatic condition caused by the indirect impact, the concussive effect, of the bullet on the ear bone, and that the chances of recovery were practically nil. But he just replied that it would take a while. With everything Serpico was undergoing at the time, Dr. Ugar felt there was nothing to be gained by telling him that in all likelihood, regardless of what else happened, he would be permanently deaf in the ear.

Serpico was still so woozy when the first card had come, the one calling him a "scumbag," that he really had not reacted to it. But the second card, with its vicious "rat bastard" accusation, left him hurt and depressed. When he had embarked on his lonely fight against corruption in the Police Department, he had believed that the blame lay with the system, not individual cops, and he had to ask himself now what kind of people could have sent him cards like that. Were they truly examples of how much he was hated?

He never ceased to be amazed at the idea that he was a "rat," a fink, an informer. It would have been different, he supposed, if he had sworn fealty to, say, even the Mafia, and then spilled its secrets. He could see that, in all of its variations. But the only oath he had taken as a cop was to uphold the law, and there was nothing in it that said that policemen had some special immunity. If anything, he thought, the opposite was true, that it was incumbent upon a cop to adhere to a stricter standard of conduct than the average citizen, to exemplify what society should be rather than reflect what it was. Perhaps that was asking too much, perhaps not; in this regard he could only answer for himself.

A few days later the door to his room opened, and a nurse started to enter. He saw her pause and turn to speak to someone outside. Then she came in, shaking her head. "What was that all about?" he asked. The nurse said that the cop on guard wanted to know what Serpico looked like, and she had told him he could come in and see for himself, and he had said, Oh, no, he couldn't, that he had been told that he couldn't. Then Serpico suddenly realized that the hospital was in the 80th Precinct, and that the patrolman guarding him must be from there. It was the same precinct where he had caught two cops shaking down his brother's grocery store and he could imagine what they were saying in the station house.

Eventually some of the younger officers did come into his room to chat with him. When the first of them walked hesitantly in and said, "Hi, how you feeling?" Serpico said, "What's the matter with you? Didn't you get the word?"

"Sure I did, but *fuck* them. I'm supposed to be a troublemaker too." He told Serpico that during a police "action" for higher wages that had taken place in the city a few months before, he had insisted on going to work anyway. As a result, he was transferred from a pleasant precinct in Queens to the less desirable 80th, which included a large chunk of the Bedford-Stuyvesant ghetto.

Their conversation inevitably turned to corruption among policemen, and Serpico was saying something about it when a nurse came in. He continued to talk, and the cop reddened and put a warning finger to his lips. Serpico ignored the signal, and went on talking. The cop got up and abruptly walked out of the room. Only after the nurse had left did he return.

"You shouldn't say things like that in front of people," he said. "They'll get the wrong idea."

"Are you kidding?" Serpico said. "You think people don't know, she doesn't know, what the hell's going on? The trouble with you is you listen to the bullshit these guys hand out. It's about time we started admitting what it's really like, and do something about it. That way we'll get the people's confidence."

The cop was unconvinced. "I don't know," he said. "We have to think of our image. We can't wash our dirty laundry in public."

That was the crux of the problem; even this young, somewhat rebellious officer clung to the idea that it was a question of "us" versus "them." Cops didn't seem to think of themselves as part of the community. Too many of them, Serpico thought, had isolated themselves not only professionally, but socially. They believed they were misunderstood by the "outside world," that there was a general public antagonism toward them, and this notion had fed on itself back and forth, until in fact there *were* "sides" and neither side could relate to the other. If cops talked to people more, instead of just other cops, if they took the initiative in reaching out to them. Serpico was certain they could break down some of these barriers. But they tended to withdraw more and more into themselves and became contemptuous of the public. Cops, after all, saw the seamy side of life every day; they knew what the public was capable of. It colored their whole outlook. If a kid had long hair, he was a sure bet to have dope on him. If a girl happened to have birth-control pills in her purse, she was promiscuous. "Oh boy," Serpico recalled hearing a cop say on one such occasion, "she puts out."

He was willing to grant that in a nation where, despite all the rhetoric, the ordinary white man couldn't relate to a black man, it was demanding a great deal to ask a white cop to assume the burden. But, he thought, there had to be a bottom line somewhere; and a cop had a social responsibility that he could not ignore. Serpico saw policemen forgetting what their job was all about, beginning to exist just for themselves, developing what amounted to their own subculture. He was sitting around with some cops in a bar once when one of them declared, "I got my gun and I'll protect me and mine," and the others nodded in agreement. "Me and mine" meant his family, and Serpico remembered him now, wryly imagining him stashing away weapons and ammunition in his home, waiting for the revolution to sweep down.

The subculture made up its own rules. If a cop was, say, a Catholic, he went to mass every Sunday because if he didn't he was liable to go to hell. But taking money did not seem to mean anything. When Serpico was first exposed to bribery and graft in the Police Department, there was talk of "clean" money and "dirty" money. Clean money meant things like traffic bribes, or payoffs to overlook gambling and prostitution. Dirty money was in narcotics. Gradually, however, the distinction blurred, and finally it was just money.

Without really being aware of it, Serpico had started to drift away from this closed policeman's world while he was still in the 81st Precinct. At first his social life was almost exclusively with policemen, but he grew bored with the endless shoptalk, the sameness of the stories and attitudes. After eight hours on the job, even though he loved the work, he yearned for new insights, different experiences. So the initial transition was easy; he decided to return to college to get a degree in sociology, and most of his free nights were taken up with classes.

A more drastic change occurred after two years in the 81st, when Serpico was transferred to the BCI, the Bureau of Criminal Identification. There had been a departmental directive that patrolmen who wanted to take a criminal identification course in finger-printing on their own time—and got good grades—would be eligible for assignment to the BCI. Serpico had applied at once, and passed with high marks. Although the BCI was essentially a record-keeping operation whose personnel was a mixture of detectives, patrolmen, trainees, and civilians, it was nominally part of the Detective Division, and was considered a "career path" toward a detective's gold shield, an idea the department did nothing to discourage. In fact this was a canard; men were constantly brought into the drudgery of BCI work under the illusion they would be promoted, and then were informed that budgetary restrictions made it impossible "at the present time."

The only real investigations were done by the latent-print unit, which dusted for fingerprints following homicides or other major crimes, and it was staffed by men who were already detectives. At first, however, Serpico believed what he had been told. And while his job was tedious—processing prints, sending copies to the FBI in Washington to see if a suspect had a record elsewhere, making a name check for a detective in the field conducting a surveillance—he did not mind, because he was convinced that if he did it well enough he would soon be in the field as a detective himself.

The BCI offices were in an annex across the street from Police Headquarters in lower Manhattan, not far from Greenwich Village. Serpico had been working there for about a year when he finally broke up with the receptionist he was going out with, and on evenings when he did not have classes—he had switched from Brooklyn College to a downtown branch of the City University of New York—he began to spend time in the Village, wandering around, sitting in coffeehouses, dropping into jazz spots. Except for a couple of vacations to Puerto Rico to lie on the beach and to brush up on his self-taught Spanish, he had up until then remained basically a Brooklyn boy.

It wasn't long before he started picking up waitresses in some of the places he frequented. But the waitresses were not like the ones in Brooklyn. Invariably they were aspiring young actresses, singers, or dancers from around the country, tiding themselves over, hoping for their big break in New York. He was enchanted with this new turn in his life. He had liked opera from the days when he was a child, listening to it in the greenhouse with his father, but he had never attended a performance until he began dating a singer. From then on he went to the opera regularly. A dancer enticed him into going to his first ballet, and he took to it at once, finding the graceful, disciplined body movements on the stage immensely appealing. Keeping his own body fit was almost a fetish with him, and he worked out often in the Police Academy gym, ran along the beach at Coney Island, biked a great deal, rode horses at a Brooklyn stable, and faithfully kept up the karate exercises he had learned in the army in Korea.

He was still living with his parents in Brooklyn, but began to rent furnished rooms in the Village for a month at a time. He let his hair grow longer. Personnel at the BCI wore civilian clothes; there were no particular rules about what was and was not acceptable, and Serpico started dressing in increasingly casual fashion, sometimes in blue jeans, even on occasion wearing sandals. The cops in the office chatted mostly about sex and sports, and his inability to do so set him further apart. He was having all the sex he could handle and felt little need to talk about it, and he had never been interested in the usual sports, save for some amateur boxing he had done in his teens.

None of this went unnoticed in the office, and eyebrows were raised even higher when he was found reading a ballet program. The confusion of his fellow officers was compounded when a stunning stewardess he had met on one of his flights to Puerto Rico came by the BCI several times to pick him up for a date. He knew what everyone had been thinking, and privately he enjoyed the discomfiture he was causing.

"Hey, Frank, that's some broad," one of the married cops said. "Where you going with her?"

"Well, Allegra Kent's dancing at City Center tonight. Maybe we'll try to catch that."

"Frank, you got to be kidding."

During a vacation he grew a full beard for the first time, and for a brief period he kept it on at work. Part of the reason was that the Village was having an influence on him, and partly it was rebellion against the dull routine of his job. But there was another thing too: every once in a while scruffy, bearded characters came into the Bureau of Criminal Identification, and he would hear someone ask, "Who's that?" and someone else would say, "He's an undercover cop," and instantly Serpico would slip into a fantasy in which he, too, was working undercover.

Perhaps he went too far one day when he stopped by the office to collect his paycheck. As a gag more than anything, he brought along a pair of toy poodles that he was walking for one of his girlfriends. The inspector in charge of the BCI, a humorless, red-faced, bucktoothed man with thinning white hair, happened to be absent, and Serpico put the poodles on his desk. Unfortunately, one of them peed on it. For a week or so the incident drew laughs around the office. Then Serpico noticed the inspector eyeing him coldly, and was certain that he had found out.

Not long after, Serpico came down from the fifth

floor, where he usually worked, to the fourth floor to file some papers. He had to go to the men's room. As he did, he recalled that several days before some of the patrolmen had been gabbing about the excellent view the fourth-floor men's-room window offered of a girl, probably a prostitute, entertaining various male visitors in her bedroom. He had meant to go down to take a look himself, but he forgot about it and then heard that the girl had drawn her blinds. When he went into the men's room, the lights were out. He saw a figure by the window. When he got closer, he saw that it was another cop, a particular favorite of the inspector's, a rather reserved young man. "You too," Serpico said. "I thought she pulled down the shades."

"Yeah," he said. "Anyway nothing's happening."

They left together. As they were going out the door, another man came in. The cop with Serpico kept going. Serpico paused momentarily to let the third man in. The lights suddenly went on, and Serpico found himself face to face with the inspector, his hand on the switch. It was a ludicrous scene, and he tried hard to keep a straight face, and then thought nothing more about it.

But as soon as he got back to his desk, a lieutenant who was second-in-command told him, "The boss wants to see you."

In the inspector's office he sensed immediately that something was very wrong. The man was flushed so deeply his whole face seemed to glow; he was quivering with rage, barely able to control himself. But Serpico was totally unprepared for the ugliness that followed.

"What were you doing there in the shit house in the dark with another man?" the inspector demanded.

"Sir?"

The inspector's lips twisted in a sneer. "You heard me," he said. "Were you sucking his cock, or were you just a Peeping Tom?"

Serpico was speechless. He could not believe his ears. Then he felt himself flushing, and the anger rising in him. "Inspector," he said, "that's pretty heavy stuff you're throwing around."

"Get that other man," the inspector snapped, "and bring him in here."

Serpico stared back at him. "No," he said, "you might think there was some collusion between us. I'll wait here, and *you* send for him."

The inspector called in the lieutenant. Then he said to Serpico, "Who was he?" When Serpico gave him the name—the name of one of his fair-haired boys—the consternation on the inspector's face was obvious. It almost seemed as if he regretted starting the whole business, but it was too late now, and he told the lieutenant, "Get him in here."

The other cop came in, looking questioningly at Serpico and the inspector. The inspector's voice was more composed but still had an edge to it. "What were you doing in the men's room in the dark with another man?"

The cop blinked. "Wait a minute, inspector! I'm a married man with three kids!"

Serpico broke in at once. "What's that got to do with it? I'm a single guy with no kids. The point is that some very strong accusations have been made here."

The inspector tapped his fingers impatiently on his desk. "Nobody's making any accusations." He turned to the other cop and told him he could go, and the cop left.

The inspector regarded Serpico for a moment, and said, "I couldn't stand you from the day you walked in here."

"I'm sorry to hear that." Serpico kept his tone as formal as he could.

"You live in the Village and you come in here with sandals on and a beard and . . ."

"Inspector! I only came in with a beard after I was on vacation, and I shaved it off in a couple of days. If you had let me know your feelings, I would have tried to comply with them."

The inspector suggested to Serpico that he ought to consider whether he still wanted to work at the BCI, or if a transfer might not be preferable.

"I certainly don't want to be transferred under these conditions," Serpico replied. "I can imagine what people will say."

"In that case, you just better toe the line."

"Inspector, I'd like to get one thing straight. Am I doing satisfactory work?"

"You're not on the bottom of the totem pole, but you either behave yourself, or I'll bite your head off."

Considering the circumstances, Serpico thought, the inspector's parting threat had interesting connotations.

Serpico was resigned to the fact that he would be transferred back—"flopped," a cop would say—to the uniformed force sooner or later. He would not miss the BCI, but it meant that he would be dropped from the Detective Division, and he worried about what that would do to his chances of ever becoming a detective. He had been hoping somehow to find a way to get on the pickpocket squad, which worked out of the same building. At the time, however, his greatest concern was to cover himself in case the inspector spread any tales about him. But it was not easy, for he had no influential contacts in the department.

The next day, Serpico sought the advice of the Catholic chaplain. He explained what had taken place. The chaplain seemed horrified. He told Serpico that it was unbelievable, that in effect the inspector had charged him with a loathsome felony and that he had every right to redress if anything further came of it. But when Serpico said the thing that troubled him the most was the possibility of rumors getting started, the chaplain replied that rumors did not mean anything. "After all," he said solemnly, "they even say things about me."

Still apprehensive, and angry, Serpico did something beyond his wildest dreams. He went out and bought a miniature wire recorder, determined to confront the inspector again and take down what he said. Once he secreted the recorder on his person, he first related the entire story to the BCI lieutenant.

"He said that?" the lieutenant said.

"That's right."

Then the lieutenant told Serpico the inspector was convinced that some sort of "hanky-panky" was going on in the men's room, that he had discovered a pair of shorts with semen on them in one of the stalls.

What the hell did he do, Serpico remembered wondering, send them to the lab for analysis?

"Listen to me, Frank, I'm not going to let you see him," the lieutenant said. "Nobody's going to get transferred. Forget the whole thing, for God's sake. Don't crank it up again."

Serpico reluctantly agreed. But within weeks he was put in a group being sent back to uniformed duty. The excuse was that room had to be made for new trainees in the Bureau of Criminal Identification. Although there was nothing in his official record, his instinct that the incident would haunt him proved correct. Five years later, when he took the witness stand against a cop who epitomized corruption in the department, it was dredged up in an attempt to discredit his testimony.

By February 24, his third week in the hospital, the drainage from Serpico's ear was reduced to a dribble. A few days before he complained of some pains in the left side of his chest, and an electrocardiogram and X rays were immediately taken. The X rays showed that he was suffering from a mild case of pleurisy, probably due to exposure to the cold the night he was shot.

He had also displayed some moments of confusion in speech and thought, and an ultrasonic test—echo encephalography—was done on him to see if there had been any brain displacement as a result of the shooting. But this turned out to be negative, and gradually the condition disappeared. To determine if there were any other indirect brain problems he was subjected to a brain scan with radioisotopes. This, too, was negative.

As Serpico continued to improve in general health and orientation, he spoke of some pain in his left cheek. Then the facial paralysis, which had twisted his mouth, began to go away. He was given a series of electrical tests to evaluate the muscles that the bullet had plunged through. They indicated that there had been only partial nerve damage, and to the naked eye at least Serpico's face began to appear quite normal. This kind of recovery was rare, and Dr. Ugar was particularly pleased by it. From time to time he would bring in neurosurgical students and visiting doctors to look at Serpico, and retrace the details of his wound and its treatment.

"What am I," Serpico asked him, "on exhibition or something?"

"I like to show you off," the doctor said with a quick smile. "You are a very lucky man."

A television set had been put into his room but he found almost everything on it jarring, and he became very critical of whatever he watched—the news, the melodramas, the commercials. They all seemed so idiotic and pretentious. In the end the only show he watched with any regularity was the children's educational program *Sesame Street*. It had a gentleness and lightness that relaxed him, and it was about all he could take.

When he was finally allowed visitors, they sometimes arrived in droves, and the room became a cross section of his life. On occasion they made an odd lot. John O'Connor, his old pal from his recruit days at the

Police Academy, would be there. So would Chief Cooper, and an inspector named Paul Delise, who had given him moral backing when he most needed it, and a handful of other cops who had remained friendly to him. Boyhood chums from his old neighborhood, who had read what had happened to him and whom he had not seen for years, stopped by to reminisce. There were young assistants from the Bronx district attorney's office who had worked with him on corruption cases, and reporters who had covered them, and neighbors from the Village, painters and actors and students. One of them, the owner of a small art gallery, brought a guitar, and everybody spent an afternoon singing folk songs. And there were the girls—stewardesses, the black model, a topless dancer. His mother came each afternoon. Usually she would sit to one side, talking to a friend who had accompanied her. Once three of them came, elderly women dressed in black, gossiping in Italian among themselves. They had all brought Serpico the same present, pajamas—"so Frank will look nice."

Suddenly, at the end of February, he went into a severe depression. A nurse told him that a cop had been killed. "It almost sounds like what happened to you," she said. "It was in the papers."

He asked her if she would bring him the clipping, and the next day she did. The man who had died was a young black patrolman named Horace Lord. He was a member of the elite PEP squad, short for Preventive Enforcement Patrol. It was a police detail made up exclusively of black and Puerto Rican policemen, and its widely heralded mandate was to patrol the Harlem ghetto on a fairly freewheeling basis to cut into illegal gambling operations, especially the numbers racket, as well as narcotics traffic, and to clear the streets of muggers, purse-snatchers, and prostitutes. Periodic bulletins were issued attesting to the PEP squad's success, listing the number of arrests it had made, the bags of heroin seized, the number of rifles and pistols confiscated. One announcement proclaimed: "The purpose of the unit, in addition to law enforcement, was to prove to slum residents that the Police Department really cares about them and can provide excellent service to the community."

According to the story Serpico read, Horace Lord had been killed during a shootout with two narcotics suspects in a Harlem tenement, and his heart sank as he went over cursory details of Lord's death.

He had met Lord not much more than a year before, when Serpico was working as a plainclothesman in Harlem. Serpico's activities against corruption in the department were common knowledge by then, and he was almost completely ostracized by his fellow cops. He was sitting in court for the arraignment of a prisoner when Lord sat down next to him, introduced himself, and told Serpico how much he admired him.

"You got a reputation with my people," he had said. "You're fair, and they respect that, even though they're scared of you. At least you don't give whitey all the breaks. You know they call you 'The Ghost,' don't you?"

The reference was to Serpico's propensity for disguises on the job, the way he moved from one rooftop to another before swooping down on a gambling operation. He had heard about the name, and he smiled and said, "Well, that's an exaggeration. Anyway, I hear you're with a pretty good outfit."

"Shit, man, you don't know the pressures," Lord said. "They're twenty-some of us, and there ain't but three that aren't already on the take, and I'm one of them. I wish I could do what you did."

Serpico was dumbfounded. He had assumed that the PEP squad was everything that had been claimed for it, that while white policemen were capable of corruption, especially in the ghettos, a select group of black and Puerto Rican cops would not prey on their own people. Finally he told Lord that they would get together and maybe talk over the problem. And now Lord was dead. It had all gone sour, Frank Serpico thought as he lay in his hospital bed. First the "condolence" cards, the cops on guard outside not speaking to him, and then Lord, who had wanted to help his community, sacrificing his life for it. Serpico had always wanted to be a policeman, and now the policeman he had so looked up to in his boyhood was the enemy whom he had to fear; in a way, as a cop himself, he had become his own enemy. He had to escape somehow, start a new life somewhere.

He remembered fearing cops as a boy, but that fear had come out of respect, and the respect wasn't there anymore. He recalled the times when he was a plainclothesman with a beard, looking like a hippie, driving his car and being stopped by a cop for no reason other than his appearance, and showing the cop his patrolman's shield and saying, "I'm on the job," and the cop always laughing in response, "Hey, that's a real good one. Glad you're working for us."

And he thought again about what would have happened without the protection of the shield. What sort of trouble would he have been in then? In the afternoon he dozed off, and had a dream. In the dream all the policemen in the city were hippies with beards and long hair and love beads. Two hippie cops stood on Madison Avenue and watched a neatly trimmed man pass by, and one of them said to the other, "How can a guy walk around with a haircut like that?" Another hippie cop stopped a well-dressed businessman driving through a red light. The businessman offered him a bribe, and the hippie cop said, "You straight people are all alike. You think you can buy yourselves out of anything."

Later, Chief Cooper came to see Serpico, and found him in a bitter, sullen mood. Serpico told him how he felt about Lord, and Cooper tried to raise his spirits. "Come on, Frank, you're reading too much into this."

Serpico ignored him. He told Cooper that he wanted the cops on guard outside his room removed. Cooper replied that they were there not to guard him, but to keep him company.

Serpico laughed derisively. "You know how many of them have come into this room since I've been here? Three! Who are you kidding?"

Cooper became angry himself. "Listen, Frank," he said. "I don't know if you know it, but thirty-five cops offered to give blood the night you were shot. One guy drove in all the way from Long Island."

Serpico had not known this, but it was too late for

him to turn back. "Isn't that great," he said. "Out of thirty-two thousand cops, I've got thirty-five friends."

Then Cooper played his trump card. He had come with the news Serpico had always waited for, the news that he would be promoted to detective. "Frank," he said, "you're going to get the gold shield. It's been approved."

"Tell them they know where they can shove it," Serpico said.

That night, alone in the hospital, Serpico cried for the first time as an adult—out of frustration and rage and sorrow.

chapter 6

Serpico's flirtation with Greenwich Village ceased temporarily, except for occasional off-duty days, when in early 1965 he was transferred out of the BCI and returned to uniformed foot patrol in the 70th Precinct, in the middle of Brooklyn. It was a considerable change from the 81st. Much of it included the western edge of Flatbush, one of the borough's best residential areas, with its rows of well-kept brick houses and tree-lined streets. There were a few apartment developments and, at the precinct's lower end, toward Coney Island, some small factories, garages, and used-car dealers. The neighborhood was generally white, mostly Jewish, with some other ethnic pockets, and middle class. If it was not precisely the country-club kind of precinct he had heard about at the Police Academy, it was quiet.

To commute there from the Village was too much trouble, so at first he rented a furnished room, then an apartment, near the precinct. A long-standing regulation of the department said that no police officer could live in the precinct where he worked, and this was one of the examples that the department's hierarchy liked to cite to show how alert it was to the danger of corruption.

When Serpico reported for duty, his hair was long by station-house standards, and he was sporting a rather large, bushy mustache.

"The first thing that goes is that fucking thing on your lip," the desk lieutenant said. "And get a haircut."

But the precinct commander, Captain Joseph Fink, called Serpico into his office. Fink was an imaginative officer who later received widespread praise for his diplomatic handling of the great hippie and flower-child migration to Manhattan's East Village. Fink told Serpico to keep his mustache on and his hair the way it was, that they were just right for some special assignments he had in mind.

But the barrage of criticism about Serpico's appearance from the precinct lieutenants and sergeants, his immediate supervisors, continued, and it helped to lay the groundwork for future stories that he was some sort of "psycho." "How can you go around looking like that?" a sergeant would say. "If I was the captain, I'd write your ass up."

"Yeah, well, you're not. The captain says it's OK."

Fink used Serpico from time to time to move around the precinct out of uniform, checking complaints that he had received from parents worried about narcotics. All Serpico's arrests involved minute amounts of marijuana, and when he learned more of the reality of the drug problem, of what was and was not important, he would regret having wasted his time. There also had been a rash of burglaries, shop windows being smashed and merchandise snatched, and when Serpico was on a midnight-to-eight tour, Fink permitted him to bring his own car to the post. He would either patrol in the car, or park it, take off his hat and uniform coat and put on a sports jacket, and casually walk the streets.

The night a burglar was caught nearly cost Serpico his life. Part of the city's subway system ran on the surface through the precinct, and a street-level station offered an excellent observation spot along an avenue where a burglary had taken place a few nights before. Inside the station, in his sports coat, Serpico saw a man standing on a corner opposite him. It was about five A.M. The man was carrying a brown paper bag, and he kept peering around. Possibly, Serpico thought, he was just someone off on a fishing trip out of Sheepshead Bay, waiting for a friend to pick him up. It was a common enough sight around the precinct at that time of the morning. But then he saw the man look up and down the intersection once more, and then quickly move away from it.

Serpico waited until he was perhaps two-thirds of a block ahead of him before he followed in the shadows. Suddenly there was the sound of glass

breaking—a jewelry-store window, it turned out—
and an alarm went off. Serpico drew his revolver and
started down the block. Ahead of him he could see the
man on the sidewalk, holding the bag. Just then
brakes squealed on the corner beyond the man. In his
unorthodox efforts to stop the burglaries, Captain
Fink had assigned two other patrolmen to cruise the
area in an unmarked car, and they happened to be
close by when the alarm rang.

The man began running directly toward Serpico.
The two patrolmen in the car jumped out and chased
after him, shouting to him to stop. He looked back,
but kept running. He was practically on top of Serpico
when one of the cops fired. Serpico heard the slug
whiz past his head, and dove immediately to the pavement. A second later the man went sprawling over
him. The paper bag burst open, spraying the sidewalk with watches and transistor radios. Serpico
lunged for the man and pinned him, and hollered to
the patrolmen, "Hold it! It's me, Serpico, you fucking
assholes."

"Jesus," one of them said, "we didn't see you." And
the second one, after he saw that Serpico was all right,
quickly asked, "Hey, can we have the collar? I fired the
shot, and I have to go through the whole thing with
ballistics, and, you know, it'll look better if we got
something to show for it."

In a situation like this, when a police officer fires
his gun, it has to be reported to the station-house
desk officer, then to the department's ballistics unit,
along with the gun's serial number, and a search of the
area has to be made to find the bullet in case of

accidental property damage or claims of personal injury. What infuriated Serpico was the cavalier way the shot had been fired without a warning call. But after cursing the two cops for their stupidity, Serpico finally gave in to their pleas and let them have credit for the arrest, although he told Fink what had actually happened. "Well, they're schmucks," Fink said, "a couple of klutzes," and he thanked Serpico for not disputing the arrest.

It was the only time during his year in the 80th Precinct that he had to draw his gun. Indeed, at times the station house was so peaceful that the arrest of a "dicky waver," a man who exposed himself, was the major law-enforcement event of the day. Still, Serpico was delighted to be out of the BCI, away from its endless records and filing cabinets, back on the street, communicating with people. In the morning, on his way to the station house when he had a day tour, he always stopped to carry an elderly Italian's chair out of the house, so that the old man could sit in the sun on the sidewalk. He got to know a woman who owned an antique store; she had a serious lung condition that required her to have oxygen equipment, and he dropped in regularly to see how she was, to let her know that the police were always nearby. A stretch of Coney Island Avenue in the precinct had a number of antique shops, and when he was on post there, he learned about furniture, what to look for, what was good and what wasn't.

And there were those moments, part of the daily life of almost every cop, that left him baffled by what human beings were capable of. One night he answered a suicide call in a private house. The suicide's father grimly let him in, and pointed upstairs. He saw a woman, the mother, rocking back and forth, sobbing, on a sofa in the living room. In a bedroom, he found the body of a boy, twenty-two, he would be told. Serpico had seen dead bodies before, but this one was bad, much of his head and brains splattered against the wall behind him. A rifle and a note lay next to the body. The note said: "Dear Mom and Dad. Please forgive me. I can't think of any other way out. Please pray for me."

Serpico wondered what could have driven the boy to so desperate an act, to write so poignant a note, and after he called the medical examiner's office, he had to question the distraught father about the details surrounding the suicide, and learned that the boy had been in love with a girl. The girl wanted him to buy a car. He purchased a used one with some money he had saved, but it was a dud, constantly in the garage for repairs, and the girl broke off with him. A few hours later he shot himself.

But there were also the other moments that filled him with satisfaction, when he was once more sure that he had chosen the most meaningful work, not just catching crooks, but helping people, even saving their lives. One afternoon he had parked his car before reporting for duty when he noticed some smoke curling out of a ground-floor window in a house across the street. He grabbed a flashlight and ran into the house. Inside the front door there were mailboxes for three apartments. The door to the first-floor apartment was not locked, and when he opened it, he saw that one wall of the kitchen to his right was in flames. Down a short corridor on the left was a cubicle with a baby in a crib. He picked up the baby, and in an adjoining room found a man asleep. For a second he thought that the man had been overcome by smoke, but he jumped up when Serpico touched him and followed him out mumbling, "What's happening, what's happening?"

The man told Serpico that the baby was his sister's, who was visiting him. Then they heard barking. "My dogs!" the man said. "My dogs are in the cellar. Give me the flashlight. I've got to get them out."

Neighbors were beginning to gather outside, and after Serpico gave the man the flashlight, he went out and left the baby with one of them, then hurried back inside and ran up the stairs. There was more smoke in the hall, and he realized at once that going up the stairs so fast was a mistake; it caused him to gulp in more air—and more smoke. He could feel the floor starting to buckle as he kicked in the door of the next apartment. Flames and smoke shot out, driving him back, making it impossible for him to go farther. Later, it turned out the apartment was empty.

He got back outside, and saw people pointing to a third-floor window, where a woman was trapped with two children. The second and third floors of the house were set back, so that there was a ledge, perhaps six or seven feet wide, over the first floor. Serpico climbed up a drainpipe to the ledge. He realized that he would need help, and one of the men watching joined him on the ledge. While the man supported him, Serpico stood on a second-floor windowsill, reached up and took the two children from the woman and handed them down to people on the ground.

The woman was straddling the windowsill above Serpico, and he could see that she was enormously fat. In the distance he heard the sound of fire engines. "Don't jump," he said, "the firemen are coming."

"No," she yelled. "I'm not staying!"

Serpico looked at her bulk and pleaded, "Wait, just a second!" He turned and gripped the other man's arms to form a cradle, and braced himself. With that the woman jumped and plunged right through the hold. Serpico thought his arms were going to come off. He and the other man were knocked down, but nobody fell off the ledge, and the fat woman escaped with a bump on her forehead and scraped knees. The firemen had arrived by then, and she was helped down a ladder.

Besides personal satisfaction, all Serpico got was enough smoke in his lungs to put him on sick leave for two weeks. He did not get the official commendation he should have received for saving the occupants of the burning house because he didn't play the old Police Department game of recommending himself for recognition. Traditionally in the department the potential recipient of an award had to pursue it aggressively—among other things, he was supposed to pay someone on the station-house clerical staff five dollars to type the proper forms extolling his initiative and courage—and by the time Serpico returned to duty he decided the hell with it.

The Police Department was organized along military lines, and Frank Serpico had always accepted the gulf between him and his superior officers. He did not have, nor did he attempt to develop, any "rabbis"—people in high places in the department—to advance his career, as so many other cops did as a matter of course. But his special assignments from Captain Fink, and Fink's appreciation of the fact that he had not created a scene over the burglary arrest, inevitably led them into a discussion about his future as a policeman.

Serpico told him that what he really wanted was to be a detective. Fink thought he would make a first-rate one, but confirmed a rumor Serpico had already heard, that there was a new departmental policy which required a patrolman to spend four years in plainclothes duty before he would be accepted into the Detective Division. To get into plainclothes, in turn, a uniformed cop had to be recommended by his precinct commander, and Captain Fink assured Serpico that he would receive his enthusiastic backing.

When Serpico appeared hesitant Fink asked him what was wrong. Serpico did not know quite how to put it to the captain. It was common knowledge in station houses throughout the city that payoffs and shakedowns went along with being in plainclothes. This attitude was so widespread that when a plainclothesman made a gambling arrest, he was expected to give the desk sergeant five dollars and the clerical man at least a dollar to expedite the paperwork, on the assumption that the plainclothesmen were growing rich on graft and ought to spread some of the money around, however small the amounts, to their less fortunate colleagues.

"Well," Serpico said finally, "you hear all these

stories and stuff about plainclothes," and he remembered how he let his voice trail off, embarrassed to say more.

Captain Fink dismissed his reservations. He advised Serpico not to worry about what he had heard, that he would get the feel of things himself and go his own way. "Frank, it's the only way you'll make detective."

For Serpico this was the overriding consideration, and the decision that followed from it completely altered his life. Fink submitted his name for plainclothes assignment, and he was summoned to Police Headquarters for a perfunctory interview by a board of inspectors. The thrust of the questions concerned his personal finances. What were his assets? How much money did he have in the bank? Had he borrowed money? Did he owe payments on a car or other purchases? Serpico told the inspectors what his savings-account balance was, and he explained that he had no debts, that he always paid cash. He was asked what restaurants he ate in. He listed several, notable for their obscurity, in Brooklyn and Greenwich Village. This produced some amused smiles. "We can take it, then," one of the inspectors said, "that you don't frequent restaurants like the Twenty-one Club."

"No sir," Serpico replied, "I've never been there."

More smiles appeared after an inspector asked him what he knew about plainclothes work, and Serpico admitted that there had been some talk that troubled him. "That was in the past," the inspector said. "It used to be that way, but it's all changed now."

On January 24, 1966, Frank Serpico was assigned to plainclothes school, or, as it was officially called, the Criminal Investigation Course. The atmosphere from the first was that of an elite group, a sense that he and the others selected for the twenty-eight-day course were their own men now—no more roll calls, no more ringing in to the station house every hour, no more standing on street corners in uniform—that they had, as one instructor told them, "made it."

Almost at once Serpico forgot his qualms about plainclothes duty as he immersed himself in an exciting new world of investigative techniques and search warrants, developing informants, the art of surveillance. He learned about the "M.O."—*modus operandi*—of bookmakers, how to recognize and break the codes and cyphers they used to disguise their operations, and how to determine the existence of such other subterfuges as the "cheesebox." It got its name because the first one discovered by law-enforcement officers was hidden in a cheesebox. A bookmaker would rent an apartment and order two telephones with different numbers. Next he would install the cheesebox, an electrical device that connected the lines of the two phones. Then he would give out one of the numbers to his bettors. When they called in on that line, the calls were automatically switched to the second number, which the bookmaker kept open by simply dialing it himself during working hours from still a third phone outside the apartment. Thus if the police tapped the line the bettors were using and raided the apartment, the bookmaker would actually be miles away from the address listed in the telephone company's records.

Great stress was put on learning about illegal

gambling, especially the policy-numbers racket, and Serpico began growing a beard again, in anticipation of the day that he would be on the ghetto streets, already imagining himself dressed like a bum, making observations, then moving in stealthily to break up a policy "drop," where the bets in a particular area were collected, or even a "bank," where they were eventually forwarded.

At the time Serpico was in plainclothes school, traffic in narcotics was covered there, but to a much lesser degree than gambling, since the Police Department had special units that were supposed to have primary responsibility for narcotics dealers, street pushers, and addicts. He remembered how the sale and possession of both marijuana and heroin received practically equal emphasis as menaces to society, although marijuana was the only drug passed around for the class to sample. "You'll note its pungent aroma," the instructor intoned pompously, while two plainclothes candidates puffing a joint in front of him whispered, "Hey, this is real good shit."

He was warned that each month ever-watchful superiors would evaluate a plainclothesman on his dependability, judgment, initiative, knowledge, job interest, loyalty, investigative performance, and arrest activity. Of these, his arrest activity was paramount. The type of arrest was what counted, not its ultimate disposition in court. Thus, for a policy arrest to qualify as a felony rather than a mere misdemeanor, the person arrested had to have at least one hundred "plays," or bets, in his possession; over and over again, Serpico would see a plainclothesman count the plays carried by someone he had searched, and, when the number fell short of the required hundred, "flake" his prisoner—add additional plays to make up the difference. "What the hell," the plainclothesman would say, "the judge'll knock it down anyway."

Plainclothes arrests were rated in importance according to subject matter, with policy numbers first and then, in descending order, bookmaking, narcotics, such miscellaneous gambling as floating dice and card games, prostitution, degenerates, liquor-law violations, down to a general miscellaneous category. As the joke went, if a plainclothesman apprehended a murderer in a homicide case, the arrest would be listed under miscellaneous unless, of course, the murderer also happened to be in possession of bookmaking slips.

When he completed his Criminal Investigation Course, Serpico was sent back to Brooklyn, assigned to the 90th Precinct plainclothes squad. Almost at once he was involved in an incident that had no connection with plainclothes work, but that fulfilled all his derring-do reveries of a cop in action.

He had recently bought a motorcycle, a Honda 350, and had taken a girl into Manhattan to see a movie. On their way back he drove over the Williamsburgh Bridge, and was headed down a broad avenue that went through the 90th Precinct. He had stopped for a red light when he heard what he first thought was a series of firecrackers going off. He looked across the avenue, and a few feet down a side street, in front of a bar and grill, he saw a flash of fire, and then three men with pistols, and a figure slumped on the sidewalk. He told the girl to get off the Honda and to hide behind one of the pillars of the elevated subway overhead. Then he wheeled across the avenue to the corner of the side street.

The shooting must have occurred very fast, he thought; several passersby were just standing there, frozen in their tracks, transfixed by the burst of gunfire. The three men were still backing across the pavement toward a gold-colored Oldsmobile sedan, its motor running, when Serpico pulled up at the corner. Suddenly two of the men darted away from the car, around the corner and along the avenue. Even if Serpico had time to draw his off-duty revolver, a snubnosed .38, there were too many people on the avenue for him to risk a shot.

As the two men were running away, the third one jumped into the Oldsmobile and took off down the side street. Serpico immediately gave chase on his motorcycle. The car careened crazily around the next corner, sped down another street for several blocks ignoring the traffic lights, turned again, cut down still another street, doubled back, and then circled three or four more blocks to shake any possible pursuit. Serpico stayed with it, about two-thirds of a block behind, not attempting to catch the car, but trying to keep it in sight until it got to wherever it was going, and he could call for help.

Then the Oldsmobile veered back on the avenue where Serpico had been when he heard the shots. He saw the car slow down; it stopped for an instant, and the two men who had fled down the avenue leaped into it. The car accelerated once more, swerved into a side street on the opposite side of the avenue, and began a new series of twists and turns at an even faster speed than before. Serpico feared that he had been spotted, and he did something he did not want to do, but he decided he had no choice, and as the Oldsmobile, tires squealing, disappeared around a corner, he switched off the Honda's headlight. Some of the side streets they were racing through had no traffic lights, only stop signs, and as Serpico roared across an intersection, a car came out of it, missing him by inches. Ahead of him he saw a figure on the street jump out of the way of the Oldsmobile, step back into the street and then scramble back again just as Serpico thought he would have to brake to avoid him. Once, while cornering, he went into a long, sickening skid, a line of parked cars loomed up in front of him, and he thought this was going to be it, but at that last moment he managed to straighten the motorcycle out.

After what seemed an eternity, the Oldsmobile's brake lights gleamed red, and it stopped for the second time, double-parking just beyond a vacant lot. Serpico cut his motor at once. The street had a slight downward incline, and that, along with his momentum, enabled him to glide noiselessly behind the Oldsmobile until, perhaps ten yards from it, he nosed in between two cars parked next to the vacant lot.

The Oldsmobile's engine was still running as he crouched down. A car door opened and he saw a man get out of the Oldsmobile and start back toward the lot. The man was loosely holding two pistols, one in each hand. As he walked across the pavement, someone in the Oldsmobile called out to him in Spanish to hurry up.

Serpico drew his revolver and lay flat on the trunk of one of the parked cars. He was sure that the men intended to dump the pistols in the lot, and he briefly considered his position. It was after midnight and the street was deserted. The man was now no more than twelve feet from him. Serpico's revolver was cocked, and he had the man dead. But there were at least two other men in the Oldsmobile, and how many more guns he did not know. On the other hand, he knew what he had, and it was not good. There were five rounds in his off-duty .38, and Serpico did not have any extra ammunition with him. He could have, of course, continued the chase, but he had enough of that. There had been too many close calls, and sooner or later his luck would run out and he would lose the car.

Serpico spoke in Spanish, keeping his voice as low as possible, hoping the occupants of the Oldsmobile would not hear him. "Police," he said. "Drop the guns." The man whirled, bringing his pistols up, and Serpico's finger tightened on the trigger. But the man seemed to be peering among the parked cars, as if he was not certain where Serpico's voice was coming from.

Suddenly the man yelled *"La policía!"* and ran toward the Oldsmobile. Serpico fired a warning shot in the air. That left him with only four rounds, but he thought it might spur somebody living on the block to call the police.

At the sound of the shot the Oldsmobile took off

before the man could reach it. The man kept running, Serpico after him. One of the pistols the man was carrying clattered onto the sidewalk. Serpico was gaining on his quarry. The man rounded the corner, lost his footing for a moment, and then started to turn with the other pistol in his hand when Serpico threw himself at him and brought him facedown. Serpico straddled him and got his arm in a judo hold, bending the man's wrist back at a pressure point that was excruciatingly painful, where the slightest additional pressure would snap it.

He felt the man go limp under him. He didn't have handcuffs, so he maintained the judo hold while he searched him and found still a third fully loaded pistol in a coat pocket. Then he let the man up, and was standing there covering him when a lady appeared, walking a dog. Serpico identified himself as a police officer and asked her to telephone for assistance.

She looked dubiously at Serpico, at his boots and dungarees, his leather motorcycle jacket, and his beard, and said, "If you're a policeman, how come you don't know there's a police phone on that lamppost?"

Serpico felt as if he were starring in a Keystone Cops comedy. "Oh," he said with what he hoped was enough mock gravity to suit the occasion, "I see it now. Thank you very much, ma'am."

Minutes later two radio cars arrived, and Serpico and his prisoner were taken back to the scene of the shooting. More radio cars were there, and Serpico gave the sergeant in charge the license number of the Oldsmobile and asked him if it was all right to take his girl home. The sergeant permitted him to leave, but told Serpico to report afterward to the headquarters of the 14th Division, which embraced a number of precincts, including the 90th. "I think the brass is going to be interested in this one," the sergeant said.

Serpico found his girl deep in conversation with a man. "Hey," he said, "can't I turn my back on you for a minute?"

"He was trying to pick me up, Frank!"

"I was only looking after her," the man protested.

"Well, what did you expect?" the girl said irritably as she got back on the motorcycle. "You leave me in the middle of the street and go tearing off like that. Why can't you get a regular job like other men? Besides, you aren't even on duty."

Serpico was suddenly very weary. The wild chase, the tension of the potential shootout, the struggle with the fugitive in the street, the girl's exasperation and lack of sympathy, had all caught up with him, and after he dropped her off he wanted badly just to go home and go to sleep.

At the 14th Division he learned that in one respect he had been fortunate. His prisoner had a record of four previous felonious-assault arrests, two of them involving guns. He also learned that the victim of the shooting was the owner of the bar and grill, and there was some talk that it was part of an organized attempt to muscle in on bars in the neighborhood.

But the reason Serpico had been ordered to report to the division was that his role in the incident fell into what the Police Department termed an "unusual occurrence." He was ushered into an office, and an inspector sitting behind a desk brusquely announced to him, "OK, we're going to have to make this an 'unusual'. This is an official inquiry." What, the inspector wanted to know, was Serpico doing in the precinct while he was off duty?

Serpico explained that he had been taking a girl home from a movie and that going through the 90th Precinct was the most direct route to her house. He could not help shaking his head as he recalled the advice he had been given while waiting for his appointment to the Police Academy—never get involved in anything when you're not on duty.

"What's the matter with you?" the inspector snapped.

"Nothing, sir, I was just thinking of something."

Next the inspector demanded the girl's name and address. Serpico supplied them, and the inspector looked at him coldly. "Are you married?"

"No, sir."

"Is the girl married?"

"No, she isn't," Serpico said sharply, fed up with this nonsense. But now the inspector, reassured that there was nothing scandalous he would have to deal with, reached across the desk and shook Serpico's hand. "Let me congratulate you. That was a good arrest."

The inspector did more than that. Serpico was recommended for, and subsequently was awarded, a commendation medal for his "alertness and intelligent police action." He felt as good as he ever had about being a cop, pleased by what he had done, doubly pleased that it was being recognized without his having to do the customary politicking for a decoration. It was a good omen, he thought, at the outset of his new career as a plainclothesman.

His euphoria was, however, short-lived.

SERPICO

chapter 7

There were approximately four hundred and fifty plainclothesmen in the Police Department, working in so-called public-morals law enforcement, when Serpico joined their ranks. In each borough they were organized on three levels. At the apex of the pyramid was a deputy chief inspector with twenty or so men directly under him. At the next level were division plainclothesmen, roughly sixteen to a division, commanded by inspectors or deputy inspectors and lieutenants. At the bottom there were the precinct plainclothes squads, supervised by sergeants.

In the 90th Precinct, the squad consisted of three plainclothesmen plus the sergeant. Neighborhoods in the precinct ranged from middle-class to poor. The area was perhaps most notable for its long-established community of rigidly orthodox Hasidic Jews, always a must stop for city and state political candidates. There were also some remnants of an earlier Italian immigration, and growing pockets of Puerto Ricans. And, along with policy numbers, a Puerto Rican variation of it, *bolipol,* was especially prevalent; bookmaking went on as it did elsewhere in the city; and there was considerable prostitution, particularly along Broadway, a pallid version of its namesake in Manhattan.

When Serpico reported for duty, his sergeant told him that he would eventually be expected to meet monthly arrest quotas, a subject that had not been mentioned in plainclothes school and that departmental spokesmen steadfastly denied existed, but the sergeant said not to worry about it for the time being, to concentrate on familiarizing himself with the precinct, that a couple of prostitute arrests would suffice until he got settled in the job.

Plainclothesmen generally worked a nine-to-five day—the busiest hours for their chief targets, policy and bookmaking—unless for one reason or another, for example to trail a suspect or to check liquor-law violations, they elected to work at night. The choice was usually up to them. There were none of the restrictions placed on uniformed patrolmen, no roll call, no set times for duty. If a plainclothesman was late for work, he would simply call the squad room and say, "I'm on my way." A plainclothesman could drink and smoke on the job, he was supposed to meld into the crowd, although Serpico, looking at most of his fellow plainclothesmen, with their clipped sideburns, thick-soled shoes, and uniformly dark suits, found it difficult to believe they would be taken for anything but cops.

He orchestrated his own appearance. If he was working a ghetto area, getting the feel of it, searching for policy operations, he would let his hair grow long, his mustache and beard equally unkempt, and would wear an old army jacket, dungarees and sandals, even a poncho, so that he seemed to be just another junkie wandering aimlessly around the streets. But for the prostitution arrests he had been ordered to make, he had his hair cut back and wore a fedora, trimmed his beard into a neat goatee, dressed in a suit and vest, occasionally put on rimless glasses, and sallied forth into the night in the guise of a nervous businessman out for a little illicit pleasure.

He experienced a profound psychological change as well. No longer was he the highly visible, uniformed cop on post. When he went into a cafeteria for a cup of coffee, or into a bar for a drink, nobody gave him a second glance, nobody knew that he had a gun and a shield. Instead of being looked at, he was now the watcher, walking in the shadows, living the cloak-and-dagger fantasies he had once only conjured up, slumped unobtrusively against a building, hunched down in a parked car, crouched on a rooftop scanning the streets.

Plainclothesmen as a rule worked with partners, but as Serpico settled into the precinct—getting a sense of the community rhythms, constantly observing what was going on around him, refining a sense for suspicious activity so that it eventually became instinctive—he preferred to operate by himself. If there was a disturbing note, it came when he stepped into the precinct or divisional plainclothes office—and it was only something he felt, not anything specific, a pervasive atmosphere of intrigue and secrecy, plainclothesmen constantly whispering among themselves in corners and stopping the moment someone got within earshot. Serpico did not know what they were talking about, nor did he attempt to find out. He was determined to go his own way.

As the new man on the squad, he was given most of the "communications" the precinct received. These were letters of complaint—a great many of them charging police graft and shakedowns, others alleging instances of crime left untouched—written by people living in the precinct. They were addressed to the White House, the Mayor, the FBI, the Police Commissioner, and so on. Although the letters invariably contained please not "to let the precinct know about this," they were in fact returned to the precinct to look into. It was small wonder, Serpico thought, that the majority of them were sent anonymously. Often months elapsed as a communication was passed down from one command to the next, but while each one required an investigation for the record, it was apparent to Serpico that they were not considered important, but as an unwanted chore for the new man to handle.

Sometimes a communication was signed, however, and one of them led Serpico to a singular policy-numbers arrest. A woman had written to the Police Commissioner that the apartment above her was being used as an after-hours club, selling liquor illegally. To make a case, several "observations" were necessary, and Serpico made enough of them to convince himself that there was substance to the woman's allegations. So one evening he went to the apartment where the club was supposed to be and tried the door, but no one was there. Perhaps he had come too early. Serpico was anxious to get to a night college class he had, but he decided to stop by the apartment of the woman who had complained. A girl, about fourteen years old, answered his knock. When he asked to see her mother, the girl led him into the kitchen. A stout woman, her back to Serpico, was busily writing at a table. He looked over her shoulder and saw that they were policy slips. "Ma'am," he said politely, "I originally came to ask you a few questions, but now you're under arrest."

"Why you picking on a poor old lady like me?" the outraged woman sputtered. "Why don't you do something about all that partying upstairs?"

"Oh, I would," Serpico said, "except fate has intervened in the form of subdivisions 974 and 975 of the penal law. They refer to what is commonly called the numbers game. Shall I quote from them?"

The woman, as it happened, was a pillar of one of the neighborhood churches; an endless number of character witnesses appeared on her behalf, and the judge finally threw the case out on the grounds that Serpico had entered the apartment under false pretenses.

Serpico nonetheless continued to follow up the communications he received, particularly when they contained allegations of police graft. "Hey, Frank," the sergeant said to him one day, "you don't have to kill yourself over these things," and Serpico replied, "Why not?"

From then on there was a certain tension, as real as it was indefinable, between him and the other men and he would always think back on this as the time it had all started, when everything began to go bad, and suddenly there was no turning back.

About a week after the exchange with the sergeant, a plainclothesman on the division level, who had once attended some college classes with Serpico, joined him in a luncheonette. After small talk about not having seen Serpico for a while, he said, "Gee, Frank, I hear you're keeping to yourself a lot, you know, not mixing with the boys."

Serpico concentrated on stirring a cup of tea.

The plainclothesman coughed delicately. "But I told the boys that I go back a long ways with you. I said they don't have to worry about you. I told them you lived in the Village and probably smoked pot and everything, and I said one thing about Frank Serpico is that he's no troublemaker. With Frank Serpico, I said, it's live and let live."

Serpico stared straight ahead, sipping his tea.

"The thing is if there's any questions you got about plainclothes, something you want to know, why just ask. I'll be glad to fill you in."

There was another awkward silence. Serpico had a fairly good idea of the kind of questions he was expected to ask, and finally he said, "No, it's OK. I don't have any questions. I'm just doing my own thing."

"That's what I told them, Frank! But, like, if you

do have any questions, you know where to find me. Right?"

This encounter took place in July 1966. It was another of the "long, hot summers" in New York, the ghettos pulsing with unrest. In Brooklyn there had already been outbreaks in the Brownsville section that threatened to spread. And near the end of July all available plainclothesmen were ordered back into uniform for temporary riot duty.

Many of them, including Serpico, were instructed to report to the 13th Division, one of the key trouble spots. Although in the event of rioting their job was to back up the regular uniformed force on the street, it was made clear to them that they were not under the normal regulations of a cop on post, such as ringing in every hour. This would simply tie up the switchboard when it might be most needed. "OK, you're all experienced men," they were told when they first turned out for duty. "You know what this is all about, and you know what the conditions are. Don't get involved in anything. Just be available if we need you. If you go anywhere, let someone know where you're at."

A few days later Serpico was in his assigned area with another officer drinking a bottle of Dr. Brown's celery tonic in a delicatessen. He glanced out the window and saw a captain on the sidewalk. "Come on," he said to the second officer, "maybe he's looking for us." The two men stepped out, and, without a word, the captain signed their memo books, giving them a "see." Serpico thought no more about it.

The next day, August 10, Serpico was on the street

when the same captain came by again and gave him another "see," just five minutes before his designated meal period. Under the loose ground rules he had been given, it was pure chance that he was not already in a restaurant when the captain showed up, and he wondered why he was being watched so carefully. A number of plainclothesmen on the riot detail had simply let it be known that if they were needed, they would be at a neighboring firehouse playing cards. Still others traveled a considerable distance to Sheepshead Bay for its famous shore dinners.

The first faint suspicions that he was somehow being singled out flashed through his mind. He tried to tell himself that it was his overwrought imagination, that he was getting a little paranoiac. But then, to his amazement and anger, a clerical man in the division told him that the captain had lodged a complaint against him the day he had been in the delicatessen. The complaint was against Serpico only, not the other officer with him.

"What the fuck for?" Serpico demanded.

The clerical man told him that the captain had stated in the complaint that he had been forced to search for Serpico for more than an hour before he found him.

Serpico left in a daze, trying to figure out what was going on. Gradually his resentment ballooned. Quite aside from the riot-duty instructions he had received, it was an unwritten rule that plainclothesmen, even in uniform, were not subject to the same restraints that governed regular, uniformed cops. What's more, a complaint from a captain could easily wind up in his

being removed from plainclothes and transferred back
to uniform patrol, effectively ending his chances of
ever becoming a detective.

For the first time Serpico, the obedient cop, re-
belled. He telephoned one of the division inspectors
and repeated what he had heard. The inspector confirmed the existence of the complaint and said that as
a matter of fact the pink slip was on his desk at that
moment.

"What's it for?"

"For being off post," the inspector said. "Do you
have anything to say for yourself?"

Barely able to control his voice, Serpico replied,
"I've got plenty to say."

"Well, let's hear it."

"No," Serpico said, "what I've got to say, I'm saving for the trial. There's more here than meets the eye.
Somebody's trying to break my chops, and I think I
know why."

"What do you mean?"

"Inspector, I'd rather not discuss it now. I'll bring
it all out in the trial, if that's what they want."

The implied threat, mostly bluff on Serpico's part,
seemed to have some effect. "Now don't go off half-
cocked," the inspector said. "Don't worry about this
too much. We'll talk about it."

The next day Serpico was off duty. He was a crack
shot and often drove up to the police shooting range
at Rodman's Neck in the Bronx to practice on his own.
For this and for private shooting matches he sometimes entered, he needed his own ammunition, and
had become friendly with a range officer, who sold it

to him at a discount. So he went to the range to work off steam as much as anything else after his conversation with the inspector, and while he was there he told the range officer what had happened. "Maybe they're trying to tell you something," the range officer said. "What the hell, Frank, you *are* different."

"Yeah," Serpico said, "I guess you're right."

When he reported back to the 13th Division for more riot duty, the inspector he had spoken to sought him out and said that he had taken care of the complaint. The inspector added that Serpico could return the favor by "saying hello" to him at Christmas.

Then, a few days later, it happened—just like that, without warning, with such stunning, unexpected swiftness that even if he had wanted to, Serpico could no longer go his own way.

Each day the plainclothesmen on uniformed riot duty were dismissed in the garage under the building housing the 13th Division offices. The garage was reserved for official use, and Serpico was walking out of it to get into his own car, parked on the street, to drive back to the 90th Precinct and change into civilian dress. A black cop came up to him and said, "Serpico, right?"

"Yeah, that's me."

"Here, I've been holding this for you," the cop said, and gave Serpico a white envelope. "It's from Jewish Max."

Jewish Max, Serpico subsequently discovered, was a well-known gambler operating in the area. "What'll I do with it?"

The cop looked blankly at him. "Do anything you want with it."

Serpico, fingering its fat heft, continued on to his car. When he got behind the wheel, he looked at the envelope again. In the upper-left-hand corner a number had been crossed out in pencil, and another number—"300"—had been written beneath it. Otherwise the envelope was unmarked. He opened it and then he saw the money, a thick wad of bills. They were all tens and twenties, all old. Later, when he had a chance to count them, he found they totaled three hundred dollars, the amount penciled on the envelope.

Serpico felt sick, sweaty. He glanced around the street guiltily, as if he were being watched, and thought, What if someone finds me with this? Three or four other plainclothesmen also going off duty were coming out of the garage and getting into their cars. None of them took any note of him.

Serpico sat there, holding the envelope. He had no idea—nor would he ever—if the money was actually for him or for delivery to the precinct squad. And he would never know if the cop had simply been stupid in handing it to him, or if the payoff system was so entrenched and so open that this kind of carelessness was normal, or if he was being tested to see what he would do.

He stuffed the envelope inside his shirt and drove back to the 90th Precinct station house, took off his uniform, and put the envelope with the money in his locker.

At first Serpico did not know where to turn. This was no traffic-ticket bribe or petty shakedown. The days when everything was just a rumor were over for him. Whether by chance or design, a confrontation that he dreaded had been thrust upon him. He had to do something—but what? Anticorruption activity in the Police Department was the direct responsibility of its number-two man, the First Deputy Commissioner; John F. Walsh, a craggy-faced, feared, remote figure.

How, he thought, could he get to Walsh? In the rigidly military fashion the department was structured, it was tantamount to a private in the army seeking an audience with the chief of staff. How many intermediaries would he have to go through; explaining why he wanted to see Walsh? After the pressures that had been grinding on him during the last few weeks—even an inspector asking for a gift at Christmas—Serpico mistrusted almost everyone around him.

His dilemma about Commissioner Walsh applied equally to the other anticorruption details in the department, notably the Police Commissioner's Confidential Investigating Unit and the Chief Inspector's Investigating Unit. There was also the small, under-manned Chief of Patrol's Investigating Unit, then headed by Sydney Cooper. If Frank Serpico had known Cooper then, or someone close to Cooper, it might have all ended differently, but he did not. He did not know anybody, and what he had to talk about required a full-scale investigation, a sympathetic ear, an official who cared. There were no witnesses to what had happened in the garage. There was the envelope with the money, of course, but after that it was, as lawyers like to say, a one-on-one situation,

Serpico's word against that of the cop who had given him the envelope, and Serpico didn't even know the cop's name.

He had to confide in someone, however, and late that night he decided at last to seek the advice of another plainclothesman who had been in the same Criminal Investigation Course with him. The plainclothesman's name was David Durk. At the time he was attached to the city's Department of Investigation, which reported directly to the Mayor and was empowered to look into the affairs of any municipal agency.

David Durk—at thirty-one about the same age as Serpico, six feet tall, with a somewhat sallow complexion, prematurely graying hair, and an aggressively articulate manner—was the son of a doctor. He was also a college graduate, not as a result of the grubby route through night classes at branches of the City University that most policemen followed, but from prestigious Amherst, where he had majored in political science. He had attended Columbia Law School for a year before deciding that a lawyer's life was not for him and had become an importer of African carvings. Still restless, and by his own account intrigued with the challenge of being a cop, Durk joined the force in 1963, four years after Serpico.

Serpico first met him when they appeared at Police Headquarters for interviews before being accepted for plainclothes school. Durk was wearing a gray flannel suit, a rep tie, and a button-down shirt. The shirt collar was a bit frayed. "It's my integrity shirt," Durk told Serpico, and they both laughed.

Serpico was drawn immediately to Durk, and the two frequently lunched together during the Criminal Investigation Course, and later Serpico occasionally had dinner with Durk and his wife. Not only had Serpico never seen a cop who dressed like Durk, but he had never known one who talked like him. They constantly discussed the problem of corruption among the police, and how it affected a policeman's role in the community, and Durk was able to verbalize easily and impressively much of what had been churning inside Serpico: that corruption flourished because highly placed officials in the department tolerated it, and how this same tainted leadership placed so little emphasis on the fact that being a cop meant serving, helping others.

But it was not Durk's confident facility with words that prompted Serpico to call him about the envelope. During their time together in plainclothes school, Durk had hinted that he had powerful contacts in the new administration of Mayor Lindsay. Often, Serpico recalled, Durk had been summoned out of class in the middle of a lecture to answer telephone calls; it was highly unusual, and Serpico reasoned that somebody important must have been at the other end of the line. If he had any doubts about this, they vanished when Durk, despite the departmental policy that plainclothesmen had to serve four years before being eligible for the Detective Division, was whisked immediately into the division and was assigned to the Department of Investigation, where he awaited his detective's gold shield.

Durk, in truth, had political connections with the Lindsay administration beyond anything Serpico then imagined. They had begun with another Amherst graduate, Jay Kriegel, a whiz-kid operative who although only in his mid-twenties was one of Mayor Lindsay's innermost circle of advisers.

According to Durk, he met Kriegel while on foot patrol near Central Park early in the summer of 1965. The two fell into conversation, discovered they were both Amherst men, and Kriegel expressed some amazement that Durk was a cop. They started seeing each other on a regular basis. John Lindsay was already running hard in his first mayoral race, and Kriegel asked Durk for some background information concerning police efficiency for use in the campaign. Durk readily obliged, and also contributed heavily to a Lindsay white paper on crime.

Through Kriegel, Durk next met Robert Price, Lindsay's campaign manager, who, after Lindsay was elected, became a deputy mayor and, next to the Mayor himself, the most powerful figure in City Hall. Among other things, Price was chief talent scout for the incoming administration, and he introduced Durk to Arnold Fraiman, Lindsay's choice for Commissioner of the Department of Investigation.

Approximately twenty men from the Detective Division were on special assignment at the Department of Investigation, and one of Fraiman's first tasks when he took office early in 1966 was to find a man to head them. He showed Durk a list of candidates. Durk looked them over and told Fraiman that the best man for the job was not on the list, a Captain Philip Foran, an officer of considerable Gaelic affability and charm who was then attached to the CIIU, the Chief Inspector's Investigating Unit. Foran, said Durk, was knowledgeable about corruption, dedicated, and had great leadership ability. At a subsequent session with Fraiman, Durk continued to press for Captain Foran, describing him as the "most honest cop I've ever met."

As would become increasingly apparent, Durk found it irresistible not to express an opinion on almost any subject when asked. Actually, for all his rave notices regarding Foran, he had not known him very long. He had met him in some classes at Bernard M. Baruch College in Manhattan, where Durk was pursuing a master's degree at night. They had talked about police corruption, and Foran was apparently impressed enough to arrange Durk's transfer into the CIIU. Durk served there only a brief period, however, before he was accepted for plainclothes training. Meanwhile his proselytizing on Captain Foran's behalf paid off and before an official approach was made to Foran to take command of the detective squad at the Department of Investigation, Durk called him and told him that if he wanted it, the job was his. Foran was incredulous that a plainclothesman was in a position to discuss such an offer, but when he realized that Durk was serious, he said that of course he wanted it. Foran was appointed to the post, and shortly thereafter Durk went to work for him. Later, when Durk was asked if his relationship with Kriegel and Deputy Mayor Price had anything to do with the assignment, he replied with a straight face, "I'd like to think not."

The day after Serpico called Durk about the envelope, they met to discuss what to do. Durk seemed excited by the opportunity created to launch a major investigation into corruption, and he told Serpico that he had just the man to go to—his boss, Philip Foran.

Serpico asked him why not go right to the top, to Commissioner Fraiman? Durk said that he felt he really did not know Fraiman well enough for such a delicate matter. Foran, he insisted, was an experienced man who would grasp the situation immediately and know exactly how to proceed from there.

Serpico had doubts. He was sure, he told Durk, that he had heard somewhere that Foran was a former plainclothesman, and that meant trouble as far as he was concerned. But Durk replied that it did not mean anything, Foran was "legit," and he trusted him.

Serpico finally agreed, and Durk set up a meeting with Foran. Serpico went to his locker and retrieved the envelope, and he and Durk went to Foran's office. When he first saw Foran, he was struck by his remarkable resemblance—except for his height, about five feet eight—to John F. Kennedy. Serpico had admired Kennedy, and he thought this was an encouraging sign.

They were seated so that Durk was between him and Foran. After a few preliminaries from Durk explaining why it was that Serpico had come to him, Serpico started to speak of the rumors he had heard about plainclothes, of his hesitancy about getting into

plainclothes work, and of the friendly plainclothesman who had so unsubtly urged him to ask "any questions" he might have about plainclothes duty.

Foran appeared astonished. He broke in to say that
while he was aware that this sort of thing used to go
on, he did not know that it was still the case.

Serpico next retraced the unjustified complaint he
had received from the captain while on riot detail, and
how, after his protest, an inspector had stopped the
complaint from going through but expected Serpico
to "say hello" to him at Christmas. And then he told
Foran what had happened in the precinct garage, how
the cop had come up to him and given him the envelope from Jewish Max with the three hundred dollars
in it.

As he spoke he took the envelope out of his pocket,
and handed it to Durk, who gave it to Foran. He remembered how Foran toyed with the envelope for a
moment, glanced into it, and handed it back.

Captain Foran leaned back in his chair and looked
at Serpico. It was rather stupid of Serpico, he remarked, to have accepted an envelope from somebody he did not know. Serpico felt the first flicker of
uneasiness. What did Foran mean? He wanted to ask
Foran how he could have been expected to discover
what was inside the envelope if he had not taken it,
but he did not get the chance. Foran kept talking, his
voice crisp, authoritative.

Serpico, he said, now had two alternatives. He
could, of course, press the issue if he desired. That
meant that Foran would have to take him to see Commissioner Fraiman, and Fraiman would "drag" him in

front of a grand jury, and word would inevitably leak out about what Serpico was doing. Foran paused. "By the time it's all over," he said, "they'll find you face-down in the East River."

He paused again. The other alternative, he said, was to forget the whole thing had ever happened.

Serpico sat there speechless. An uncomfortable warmth flooded through his body. He was terribly confused and embarrassed. He could not believe what he had just heard from the man Durk had assured him was *the* man to lead an investigation. Was he going mad? Foran could not possibly have said what Serpico heard. But there Foran was, not more than six feet away, gazing calmly across the desk at him, waiting for an answer.

All Serpico wanted to do was to get out of the office as fast as he could. Almost mechanically he said, "OK, I guess I'll forget about it. But what'll I do with the envelope? I don't want anyone to think I kept it."

"It's up to you," he recalled Foran saying.

The conversation had become unreal, the whole purpose of the meeting forgotten. Serpico was still seated, but his head spun, as if he were suffering from vertigo. "Maybe I'll give it to my sergeant," he said at last.

Foran agreed that was a good idea. His tone was soothing, fatherly. "Give it to your sergeant," he said. "Get rid of it."

PART
TWO

PART
——————
TWO

chapter 8

The New York Police Department was organized in 1844. It is the oldest municipal police force in the United States, and graft and corruption have been endemic in its ranks almost from the start, the instances of payoffs and shakedowns through the years so common that many were not considered worth covering in the press.

But every so often a scandal occurred that nobody could ignore. One of the first took place in the 1890s as the result of the immensely profitable police protection of the city's whorehouses, and it led to the resignation of the department's chief, a major reorganization of the force, and the appointment of Theodore Roosevelt as a Police Commissioner. Roosevelt approached the job with his usual bravado. "I have," he wrote in a letter, "the most important and the most corrupt department in New York on my hands [but] I am glad I undertook it; for it is man's work." Two years later, having instituted some reforms, he had a lot of second thoughts. "I have done nearly all I can with the police under the present law; and now I should rather welcome being legislated out of office." Fortunately for Roosevelt, the country was on the brink of war with Spain, and he rushed off with considerable relief to become Secretary of the Navy.

In the early 1930s, as the Prohibition era drew to a close, an investigation into corruption in the city's government and courts revealed, among other things, that whole platoons of cops were on the payroll of such notorious racketeers as Charles (Lucky) Luciano and Dutch Schultz, and this exposé launched the political career of a hitherto obscure young prosecutor named Thomas E. Dewey, who went on to be Governor of the state for twelve years and twice the Republican candidate for the Presidency.

Then in 1950 came the Gross scandal. Harry Gross was a Brooklyn bookmaker who had been paying a million dollars annually to the police to safeguard his twenty-million-dollar-a-year gambling empire. Before the affair was over, the Police Commissioner had resigned, most of the department's high command was implicated, more than a hundred policemen, including inspectors and captains, were dismissed or quit under fire, and three of them committed suicide.

Lesser scandals, meanwhile, continued to crop up regularly. In 1959, for example, a retired New York police sergeant was involved in an upstate automobile accident; it seemed like just another case of drunken driving until $19,495 in cash was found in his car, along with a list of bookmakers from whom he had been collecting payoff money for crooked cops. And in 1964 still another instance of police on the take surfaced, also centered around gambling graft, and it reached right into, and resulted in the dismantling of, one of the department's most widely acclaimed watchdog operations: the forty-eight-man Chief Inspector's Confidential Unit.

This situation was not unique to New York. During the past quarter century scandals have repeatedly shaken police departments around the country, in Atlanta, Baltimore, San Francisco, Philadelphia, in Newark, Louisville, Reno, Kansas City, and Detroit, Reading, and Albany, N.Y.

In the summer of 1966—just about the time Serpico was meeting with Captain Foran—a study of police corruption in Boston, Washington, D.C., and Chicago showed that one out of every five policemen "was observed in a criminal violation," *even though they knew they were being watched.* The study, conducted by Dr. Albert J. Reiss of Yale University under a grant from the President's Crime Commission, involved thirty-six trained observers reporting on the activities of 597 cops chosen at random in the three cities. Crimes that were witnessed embraced actual theft, receiving protection payoffs, and accepting money to alter sworn testimony. They did not include such examples of petty graft as free meals and drinks, small gifts, and discounts on purchases.

In New Orleans, in the early 1950s, some thirty

police officers were charged with bribery and conspiracy to protect organized gambling and vice. After long prosecution delays, the indictments were dismissed on technical grounds. Then in 1969 the department was further disgraced, this time because of a police-operated burglary ring.

As a result of a federal grand jury investigation in Seattle, nearly a hundred cops—about 10 percent of the force—were tied into a carefully structured shakedown system. The city's Assistant Chief of Police was among those who were jailed.

In Denver, despite its reputation as a "clean" place to live, the existence of two good newspapers, and a presumably well-informed population, the biggest criminal operation in the city's history did as it pleased for fifteen years under tight police leadership. Denver cops turned out not only to be expert burglars, but to belong to at least five safecracking gangs before their busy sideline was finally discovered, and more than forty of them were rounded up. The bitter joke in Denver was about a shopowner who called the police to report a burglary, and was told, "We know all about it—we were there."

Over the years Chicago has been famous for corruption, and a new round of unpleasant headlines hit the city in 1960 when it was learned that a police lieutenant had accompanied the head of Chicago's Mafia on a month's vacation in Europe. Then a professional thief, facing a twenty-year prison sentence, decided to talk about other members of his burglary ring—they were all Chicago cops. It was, Mayor Richard J. Daley said with some hyperbole, "the most shocking and disgraceful incident in the history of the Chicago Police Department." Since it was also an election year, Daley brought in a new Superintendent of Police with impeccable credentials to clean house. The new superintendent reorganized and modernized the department, but behind the facade nothing basically changed. In 1967, for example, a raid on a single policy-numbers operation revealed payoff lists for ten of Chicago's twenty-one police districts. Soon afterward the chief of the department's Intelligence Division, who conducted the raid, was removed from his job. And in 1971 yet another wave of scandals—with federal grand jury probes into selling heroin, gambling protection, shaking down businesses, buying promotions—rocked the police department that Mayor Daley continued to hail as "the finest in the nation."

Incidents like these are only a sampling of the extent of police corruption across the country. But large or small, they have one thing in common. The ordinary police officer, however honest, looked the other way, or got out, or once in a great while anonymously reported a specific payoff. In none of them was there a cop willing to blow the whistle on his own, and then step forward, as would be expected in any other criminal proceeding, and testify openly in court. In none of them was there a Frank Serpico.

After his meeting with Captain Foran, Serpico did not see the sergeant in the 90th Precinct for several days because of the interruption in his normal plainclothes routine brought on by the summer riot detail in the Brooklyn ghettos. When he did, he asked if he could speak to him privately for a minute. The two men went into the dingy plainclothes locker room across the courtyard behind the main part of the station house.

Serpico took the envelope with the three hundred dollars out of his locker, told the sergeant that one of the men at the division had given it to him with no particular instructions about what he was to do with it, and so he was now passing it on. "I don't need it," Serpico said. "I don't want to get involved."

"OK, Frank," the sergeant said. "I understand."

Serpico watched as he opened the envelope, took the money out, and put it in his pocket, and that was the last Serpico saw of it. The sergeant did not seem in the least surprised; he did not ask anything more about the circumstances under which Serpico had gotten the envelope, or any details about the cop who had handed it to him.

He gave it to the sergeant because he did not know what else to do or where to turn, at that point, and he needed time, he told himself, to get his head together. More than a week had gone by since he had received the envelope and he had counted on the general confusion triggered by the riot duty to cover him while he saw Foran, fully expecting Foran to sketch out a plan of action. But when Foran failed him, he was on his own, and if, as he believed likely, the envelope was a test, questions would now be forthcoming. He could imagine someone going over to the 13th Division and asking, "Where's the money?" and someone else saying, "Serpico's got it." He had never felt so lonely in his life, and it was just the beginning.

When he had left Foran's office, David Durk had followed him, saying, "I'll see you, I'll call you, I'll talk to you." Durk did call him. He told Serpico that he had not realized the kind of man Foran was. "That fuck," he said, "he's afraid of getting involved."

"Yeah, sure," Serpico replied, "forget it."

Durk wanted to meet to plot their next move, but Serpico put him off. He did not blame Durk for what had happened, but much of his confidence in him had evaporated, and he felt that underneath Durk's so-phisticated veneer he was at best naive. It was a qual-ity, as Serpico wryly acknowledged, that he could also ascribe to himself. At least he was ready to admit it, and this perhaps was the difference between them.

He could, of course, quit the Police Department, and he considered and discarded the idea. It would be the same as walking out on his life, he thought, and he was unwilling to do that, unwilling to relinquish the concept of being a police officer that he had car-ried with him since boyhood. He even found himself clutching at the possibility—knowing all the while how ridiculous it was—that the whole business with the envelope was an isolated incident, peculiar to the 90th Precinct or the 13th Division, that he had sim-ply run across a rotten apple or two in the barrel, as one was bound to. He was not ready to concede that the barrel itself might be rotten.

There was one tangible result in turning over the money to the sergeant. Almost immediately the ten-sion he had experienced between himself and the other plainclothesmen disappeared. The reason, he thought, was clear enough. He had demonstrated to their satisfaction that he was safe, that he was taking the classic path of the honest cop who closed his eyes to what was going on. They had nothing to fear from him, and it meant more cash for them to divide up.

Serpico continued on riot duty, and when that was over and he returned to the precinct, he worked by himself most of the time. No effort was made to have him coordinate his activities with the other men. This, too, fit the pattern. Nobody wanted him around while graft was being collected, or when someone who was paying off happened by.

Then the tension came back. Instead of taking it easy, and going along, Serpico began making gambling arrests without letting anyone else in plainclothes know what he was doing or planning. Prowling the streets in his customary manner he had spotted heavy "action" in a neighborhood bar— people moving constantly in and out of it, staying long enough to make a bet, but not long enough for a drink. He kept the bar under observation for several days, and one afternoon went in when there were only a few customers. Dressed in faded dungarees and an old army jacket, Serpico was an inconspicuous figure, and the bartender, after drawing him a beer, promptly returned to a conversation he was having with another patron. At the far end of the bar, in the rear, Serpico saw a man intently calculating with a pencil and paper. He finished the beer, walked to the bathroom, and when he came back, he asked for a second beer, positioning himself much closer to the man with the pencil. The man glanced at him, and bent over his work again.

Serpico could now see that he was writing policy-number slips. The paper used for policy is extremely thin, and an adept numbers writer can make it vanish in an instant, rolling it into spitballs and flicking them away, or swallowing them if he has to. Serpico suddenly leaned over and clamped his hand on the man's wrist. He had moved so fast that some of the slips were still in view on the bar. He tightened his hold on the man's wrist until the man, recovering from his surprise, shrugged his shoulders and relaxed. Serpico lifted his hand and confiscated the rest of the slips. He ordered the four or five customers in the bar to get out, and told the astonished bartender to lock the door. Then he called the station house and requested a radio car.

Without checking with anybody, Serpico brought in another numbers arrest, then a third. Making the arrests was his job, but he also realized that he was pushing for a confrontation, that sooner or later he would hit a gambling operation that was being protected. The attitude of the other plainclothesmen became increasingly hostile, and he knew that he would have to get out of the precinct. It was not that he was afraid of a fight, he would even welcome it, but just being there sickened him.

He heard that the Narcotics Bureau was looking for men, a hundred of them. He let his beard grow for the better part of a week, then put on the scruffiest clothes he could find and went to the bureau's offices in the 1st Precinct at the lower tip of Manhattan. His idea

was to show how effective he could be as an undercover man, and he felt that he had made the right decision when he entered the office and saw several
men, neatly trimmed, in white shirts and ties, filling
out application forms.

He gave his name to the receptionist and, without
further identifying himself, asked to see an inspector.
He sensed people looking at him as if he were an informant, and got another quizzical look when the inspector received him.

"I'm on the job," Serpico said, producing his shield.
"I hear you need men, so here I am."

The inspector was dumbfounded to learn that Serpico was a police officer. "We can sure use you," he
said, and actually took out an application form and
started filling it in as he asked Serpico questions.

"Where are you working now?"

"Well, I'm in plainclothes."

The inspector's hand hesitated. "Um," he said, "I
don't know if we can get you out of plainclothes." Perhaps he was being paranoid, Serpico thought, but he
was sure he heard the enthusiasm drain out of the inspector's voice. It was as though by admitting that he
was a plainclothesman he had somehow tarred himself. He raged inwardly at the hypocrisy of the department. On the one hand he was supposed to be in
plainclothes for four years before he was eligible for
detective work, and on the other hand the mere fact
that he was a plainclothesman seemed to make him
an immediate object of suspicion. He could see it in
the inspector's face: plainclothesmen took gambling

and vice money, and it was a relatively minor step, the temptations indeed greater, to taking junk money.

The inspector completed the form and told Serpico to telephone back in a week if he had not heard from him. Serpico waited the week. "I'm sorry," the inspector said, "we couldn't swing it."

While he brooded over his next move, the decision was abruptly taken out of his hands. In November 1966—a little more than three months after he had been given the envelope in the garage—Serpico was notified that he was being transferred to another plainclothes group, the 7th Division in the Bronx. The transfer order had come so unexpectedly that he did not know—nor would he ever—if it was a haphazard event in the normal bureaucracy of the department, or whether it was a maneuver to get him out of the 90th Precinct.

Then the next day, an off-duty day, Serpico received a phone call at home from the department's communications unit offering him still another job. When he had been at the Bureau of Criminal Identification, he had periodically been pressed into service as an interpreter during the morning lineups at police headquarters because of his fluency in Spanish. The caller from the communications unit explained that a new section was being set up to handle complaints from Spanish-speaking people, and he wanted to know if Serpico was interested in working there.

"Well, I don't know," Serpico replied. "I'm in plainclothes, you know, and they say it's hard to get out of it."

"Don't worry about that. If you want it, you've got it."

"I'll let you know tomorrow," Serpico said.

The job in the communications unit—basically one of answering the telephone—was far removed from the kind of police work Serpico had dreamed about as a boy, but in his anxiety to leave the 90th Precinct almost anything seemed attractive. And what worried him about the 7th Division was that it was still plainclothes duty, and he might wind up in precisely the same situation from which he was trying to escape.

At least he now had a choice, and that night he finally decided to seek the counsel of an experienced officer in the department, Captain Cornelius J. Behan, a tall, thin-faced, solemn man who had formerly headed the pickpocket and confidence squad. A school for pickpockets in Bogotá, Colombia, regularly sent its graduates to plague New York City and when Serpico was at the BCI, Behan borrowed him on three or four special assignments because he spoke Spanish. Behan was subsequently assigned to an administrative post in the public-morals section of the Chief Inspector's office, and Serpico had recently bumped into him again while they were both studying for college degrees at night. When Behan asked him how he liked plainclothes, he had indicated that he was upset with the corruption he was encountering, and Behan appeared to listen sympathetically. Beyond this there was another factor Serpico had weighed. The captain regularly conducted weekend retreats for Catholic cops in a monastery, retreats which were devoted to prayer, meditation, and spiritual discussion; and while Serpico had never gone on one of them, he figured Behan was a man he could trust.

He called Behan at his home in Queens and explained the career decision he was facing. Without going into details, he reiterated his unhappiness with his present situation and asked Behan if he could find out what the 7th Division was like. If it was at all similar to what he had been experiencing, he would take the job with the communications unit.

Behan told Serpico that he had contacted the right man. One of his neighbors right across the street, Phil Sheridan, was a deputy inspector in the 7th Division, and he would get in touch with him immediately. Behan telephoned Serpico back the same night. He had talked to Sheridan, he said, and had asked if he could use a hardworking officer who spoke Spanish and wore a beard, and who would make a first-rate undercover man. Sheridan had replied that he certainly could. Behan said that he then informed Sheridan that the man he was referring to was about to be sent to the 7th Division but was "very concerned" about corruption. There was nothing to worry about on that count, Behan quoted Sheridan as saying—the 7th Division was "as clean as a hound's tooth."

On the basis of this conversation, Serpico turned down the communications job. He had accumulated nearly a month's vacation time and decided that he had to get out of the city for a while. In the Village he had met and had an affair with a Finnish girl who was studying ballet. When the girl left, she had invited

Serpico to come to Helsinki to visit her, and he had
said sure, thinking nothing would ever come of it. But
now Helsinki seemed ideal; it was about as far away
from New York and what he had been through as he
could imagine, and he took the girl up on her offer.

Serpico had a marvelous time in Finland. The girl
was delighted to see him. He loved the snowy, forested
Finnish countryside, the saunas, the peaceful life. He
even got a kick out of the standard tourist trip to the
Russian border. He introduced himself to Helsinki
policemen on the street, and found them warm and
friendly, and he liked the way they stepped back and
saluted smartly, not only to him but to ordinary citizens. The girl took him to a student club that featured
an American folksinger. The two of them hit it off, and
occasionally Serpico would sing with him. One won-
derfully giddy night, the audience requested "El Ran-
cho Grande." The singer asked Serpico if he knew the
words, and when Serpico said he didn't, the singer
said, "Well, we'll fake it, they won't know the difference," and they sang endless made-up choruses while
everyone roared and cheered.

To top off his trip, Serpico wanted to visit Stockholm. The folksinger, who lived there when he was
not on tour, offered him the use of his empty apartment. Serpico got lost trying to find the address, so
he went into a boutique to ask directions. With his
usual luck he found a stunning girl inside who spoke
English and turned out to be the boutique's owner.
They began talking, the talk stretched into dinner,
and he wound up with a companion for his stay in
Stockholm. She called Serpico "my Ol' Man River,"

and when he asked why, she laughed and said it was because of the way he had rolled into her life and would undoubtedly roll out again.

In mid-December Serpico returned to New York greatly refreshed, ready for a new start in the 7th Division.

chapter 9

Rule one of the mimeographed sheet of instructions to 7th Division plainclothesmen exhorted them to be "particularly careful of their personal behavior, and conduct themselves in a manner above reproach."

The division encompassed four precincts in the South Bronx, a seven-square-mile area with approximately half a million inhabitants. Its most famous landmark was Yankee Stadium. Its deteriorating middle-class neighborhoods were predominantly Jewish and Italian, and it also had vast, teeming stretches of Puerto Rican and black slums, as bad as anything that could be found in the city, or, for that matter, anywhere. Law enforcement, and respect for the law, was a joke.

A deputy chief inspector was in command of all the plainclothes and uniformed men in the division. His alter ego was the deputy inspector, Philip Sheridan, whom Behan had contacted on Serpico's behalf. There was also another deputy inspector, a public-morals specialist, who along with two lieutenants was in direct charge of the division plainclothesmen, sixteen in number, and also supervised the activity of those at the precinct level.

The division plainclothes office was a large room, painted the usual green, on the second floor of the 48th Precinct station house. There were several battered desks for the men, and a larger one for the lieutenants in another corner. The decor was completed by a blackboard with the name of each man and his monthly arrest activity chalked up on it, and a row of mailboxes. A door on the left led to the offices of the division commander and the two deputy inspectors.

Serpico arrived with little fanfare. It was the Christmas holiday season, and very few plainclothesmen in the division seemed to be around. A lieutenant told him he would be assigned a partner. There was an arrest quota, the lieutenant said—two a month per man, or four for a team, however it was divided. Serpico was given a mailbox key, a list of "k.g.'s"—known gamblers within the division's boundaries—and a cardboard 7th Division sign that said "OFFICIAL BUSINESS" to put in his car windshield when he parked. One of the men showed him where the coffee room was, and told him there was a three-dollar monthly charge for coffee and cake. The only kind of coffee Serpico liked was either espresso or a strong Turkish roast, and he said, "I drink tea, and I'll bring my own bags." He heard someone behind him—it sounded like the lieutenant—say, "I never trust a cop who drinks tea." Serpico did not know how to take it, and he decided to let it pass. As a change of pace from the cigars he usually smoked, he had just lit a beautifully carved meerschaum pipe he found in an antique shop and had reconditioned, and the same voice said, "Huh, and he smokes a pipe, too." Serpico turned to see who was doing the talking, but nobody looked up.

The next day in the division office he ran into an old acquaintance named Robert Stanard who had been a plainclothesman in the 70th Precinct in Brooklyn when Serpico was a uniformed patrolman under Captain Fink. Stanard was a husky, square-faced man, twenty-nine years old, who wore horn-rimmed glasses and was nicknamed "Clark Kent" by the other cops. He affected a brash, worldly manner, and had a tough-guy habit of talking out of the side of his mouth. He was a graduate of St. John's University, and later Serpico thought of him with grim amusement when he read that the way to upgrade the Police Department and wipe out corruption was to recruit more college men.

In the 7th Division office, Stanard clapped Serpico jovially on the shoulder. He liked to spice his dialogue with bits of Italian slang he had picked up in the streets, and, using a Sicilian diminutive of Frank's name to greet him, said, "Hey, Cheech, how you doing? I heard you were coming up here."

Serpico remembered him quite well. In the 70th Precinct they had had three or four conversations.

Stanard used to talk about some of his plainclothes details in the bars and motels around Sheepshead Bay. "Hey, *mamma mia*," Stanard would exclaim, "you won't believe the broads that hang out there!" And after Captain Fink had recommended him for plainclothes duty, Serpico had asked Stanard what it was like, and Stanard, punching him playfully in the ribs, had replied, "It's a real good deal."

Since New Year's Day 1967 fell on a Sunday, the following Monday was a holiday, and as the new man in the division, Serpico pulled duty. When he got to the office that morning he was somewhat surprised to find Stanard waiting for him. Once again Stanard greeted him effusively, and told Serpico how pleased he was that he had been assigned to the division. "You'll like it here," he said. "The setup's the best. Nobody bothers you." Then Stanard invited Serpico to go along with him. He had to check a complaint, and if it panned out, Serpico could have the arrest and get on the sheet.

Stanard drove Serpico to a place under the Jerome Avenue elevated subway tracks north of Yankee Stadium called Otto's Bar and Grill. There was a restaurant area on one side, and on the other an oval bar. Despite the hour—around eleven-thirty A.M.—a number of customers were already at the bar. Stanard ordered beers for Serpico and himself. A television set overhead was tuned in to the Rose Bowl Parade. To their left, where the bar curved, a squat, swarthy man with thinning hair, dressed in a baggy brown suit, sat on a bar stool, a cup of coffee in front of him. His name, Serpico would learn, was Pasquale Trozzo.

Trozzo's position at the bar put him within a few steps of three pay telephones. A number of men came in from the street and went directly to Trozzo. After a whispered conversation, Trozzo would scribble on a piece of paper, and sometimes the man would leave right away, but most of them stayed for a drink or two. One of the pay phones rang regularly, and Trozzo always answered it, scribbling more notations on the paper. Occasionally he left the bar to make a call himself. He was obviously taking heavy betting action, and Serpico was amazed that he was doing it so openly. He wondered what the action was about and then he remembered that all the major college football bowl games were being played that afternoon and evening.

As he watched Trozzo, he heard Stanard mutter angrily under his breath once or twice. Finally Stanard downed his second beer, wiped his mouth with his hand, and strode toward Trozzo. Serpico followed right behind him. Stanard grabbed Trozzo by the arm and snarled in bully-boy fashion, "Goddamnit, we told you to stay the fuck out of here. You were told this place was hot. We got a complaint, and now you're going to have to go."

Serpico noticed that Stanard had not even bothered to identify himself, but that Trozzo knew exactly what was going on. He seemed flabbergasted at Stanard's outburst. "Honest to God," he said, "I didn't know. Nobody told me."

"Don't hand me that crap," Stanard said. "Come on."

Trozzo looked quickly at Serpico, and then said to Stanard, "Listen, I'll give you a C-note apiece."

"Let's go," Stanard said.

"Hey, OK," Trozzo replied, "four hundred. It's all I got on me. Honest."

Stanard released his hold on Trozzo's arm. "I already told you we got a complaint and you're going to have to go, but I'll give you a light one. That's the best I can do."

"A light one" meant an arrest affidavit prepared in such a way that a defense lawyer could easily pick holes in it and get the case thrown out. As soon as he heard this, Serpico moved away from the two men. Instinctively it was something he did not want to know about, and he watched them from a distance, Trozzo gesturing rapidly, Stanard shaking his head, while the inane Rose Bowl Parade commentary drowned out what they were saying.

Finally Stanard jerked his head toward the door, and Trozzo waddled reluctantly after him. On the sidewalk Trozzo continued to protest that he had received no warning about taking bets in the bar, and Stanard said, "Bullshit. You were told, and you only get told once."

"Jeez," Trozzo said, "the wife's going to be making dinner. I wanted to have dinner with the wife and kids."

"When you get to the precinct, you can call her and tell her you'll be late," Stanard replied. "If you stop screwing around, you'll make night court and you'll be out."

Trozzo got into the backseat of the car, and Stanard

drove to the station house. During the trip he shook his head again and muttered in Italian, "What a fucking cucumber." By then Trozzo had lapsed into a sullen silence, and did not reply.

At the station house Serpico let Stanard handle the booking of the prisoner, and drifted off. He spent the rest of the day alone going through the motions of familiarizing himself with some of the neighborhoods of the division, trying to put out of his mind the exchange between Stanard and Trozzo—knowing that he could not—and wondering what would come next in a division he had been assured was "as clean as a hound's tooth."

He found out soon enough. In the morning Stanard was again in the office when he got there. He took Serpico aside and started to hand him a folded one-hundred-dollar bill. "That fuck," Stanard said. "He only came up with two bills because I booked him."

Serpico refused the money. "No," he said, "it was your collar, so you keep it. I don't want it."

"You're sure?"

"Yeah."

Stanard looked at him thoughtfully, and after a moment he said, "Cheech, let's take a ride. I want to talk to you."

Serpico half expected Stanard's offer to split the cash he had received from Trozzo. That was customary in the peculiar code of a crooked cop, and Serpico remembered the first traffic-ticket bribe he had been offered when he was a rookie, and how his partner had said, "You know, we split fifty-fifty." But he was totally unprepared for what followed.

Stanard swung away from the curb, and almost idly asked, "Hey, how much time you got on the job?"

"Six, seven years, around that."

"And so now you're in plainclothes. You know what plainclothes is all about?"

Serpico decided to keep his answer as vague as possible. "Yeah. Some things I hear, and some things I guess for myself." He took out his meerschaum and lit it.

"Christ, roll down the window a little," Stanard said. "How can you smoke that thing?"

"It relaxes me."

"Anyway, the guys were saying that . . . look, let me tell you something. You have an opportunity to make some easy, clean money."

"What do you mean?"

"I mean, you can make eight hundred a month, just like that."

"Just like that?"

"Yeah, that's what I said. *I'm* saying you can. I—" Stanard seemed to have second thoughts about what he was going to say. He drove perhaps the length of a block before he spoke again, his voice lower, less assertive. "We got a call about you from someone—I don't know who, and they didn't even have your name right, 'Sertco' I think they called you—and it was that you couldn't be trusted."

Serpico allowed himself a caustic laugh. "What were they bitching about? That I didn't like to take money?"

"Yeah, something like that. But it doesn't matter. I told the guys that I knew you from the Seven-oh, and that you were OK. What we do is check a guy out with the other cops he hangs around with, but you're kind of weird. You don't hang out with cops, so there's nobody we can really talk to. So I told the guys you were OK in the Seven-oh, and when plainclothes came up for you, you asked me about it and I told you to grab it. I told the guys, 'Yeah, he's weird and has a beard and he's around the Village a lot and stuff, but he's all right.' I mean, I told them you wouldn't hurt another cop."

Serpico stared ahead, puffing on his pipe. Out of the corner of his eye he saw Stanard glance at him. "Would you hurt another cop?"

"I don't know what you mean, 'hurt another cop.' "

"You know. Would *you* do anything to hurt another cop?"

"Well, that depends on what he's doing."

"Hey," Stanard said, "nobody does anything bad here. The only thing we do is we make a little clean money off gambling. They're going to operate anyway, and they give us money so we don't bother them."

As Stanard drove slowly through the garbage-littered streets, past block after block of broken-down, rat-infested housing, in the middle of an area with one of the highest crime rates in the city, he continued breezily, "You can't get into trouble or anything. We don't go overboard on protection here. Like the niggers and the spics, they're not covered for telephones, because they're so fucking dumb and they'll say something, and get themselves into trouble. Now the Italians, of course, they're different. They're on top, they run the show, and they're very reliable, and they can do whatever they want. But for the rest of them, like, if they're taking action on the phone, we'll bust them. And there's only certain locations they can work. If they work other locations than what we tell them is OK, we can bust them for that. And if there's any heat on to make an arrest, we make the arrest, and that's it. So we're always covered. See what I mean?"

Serpico grunted noncommittally. He could hardly believe his ears, that Stanard would be so indiscreet about the existence of a "pad"—as the systematized police payoffs were called—that he could be that stupid.

But then he realized that it was not a question of stupidity at all. It was far worse than that. Graft and corruption had become so entrenched in the department, so completely a way of life for a cop like Stanard that he could not conceive of another policeman's questioning what he did, much less doing anything about it. The only thing that concerned him was whether or not Serpico wanted to participate in the pad.

"Look," Stanard said, "if you don't want to work here, we can get you transferred. We've got some connections downtown, and all we have to do is put up a little bread and you're out. It's up to you, Cheech. You want to work here?"

Serpico was curious as to what else Stanard would reveal. "Sure I want to work here. Why shouldn't I?"

There was, Stanard grandly pointed out, another option that Serpico could elect—the Times Square prostitution detail, the so-called pussy posse, which was staffed on a rotating basis by plainclothesmen throughout the city. Stanard told Serpico that he could arrange to have him on the pussy posse for six months, and Serpico would still collect his "nut," or share, of the 7th Division pad.

"Like I said," Stanard went on, "right now it's running about eight hundred a month." Then, as if he were an executive explaining corporate benefits to a newly hired employee, he added, "This is how it works. You don't get anything the first month and a half until everybody is satisfied you're OK. But you get it back at the other end. Like, if you get transferred, you get paid for a month and a half after you leave."

Stanard drove on in silence. After a minute he said, "Well, how do you feel? You in or out?"

Serpico had to sort out his mind. Nothing would ever really match the shock and disappointment he had experienced at the meeting with Captain Foran, and he had known all along, he supposed, that he would eventually be confronted with the situation Stanard had described. The astonishing thing about it was the casual, matter-of-fact way it had come, and the one thought that had seized him now, in the car with Stanard, was how this could operate so openly without the acquiescence at least of superior officers in the department. "I don't care what you guys do," he said at last, "just so I don't get into trouble."

His answer seemed to please Stanard. "OK, fine," he said. "Boy, after all that talking, I need a drink." He turned down a street in an Italian neighborhood and parked in front of a bar called the Sportsman's Lounge. "Come on," he said, "let's go in here for a minute."

At the bar Stanard ordered a Scotch for himself, and asked Serpico what he wanted.

"Oh, a beer'll be OK."

Stanard drank his Scotch, had another, and then he told Serpico, "Wait here a second," and went into a back room. He reappeared shortly at the door of the room, called out, "Hey, Cheech," and motioned Serpico over.

At a table in the back room sat a man, in his late thirties, who looked as if he had seen too many gangster movies. His black hair was sleekly combed, and his face had a pampered, barbershop sheen to it. He was dressed in a white-on-white shirt, a blue silk tie, and a well-tailored dark-blue suit. He had large gold cuff links, and a diamond ring flashed on his left pinky finger. Serpico would later spot his photograph in the 7th Division gambling files. His name was Nino Ribustello. All Stanard said was, "Hey, Nino, I want you to meet a friend of mine, Frank Serpico. He's one of the boys."

Ribustello surveyed Serpico, his beard and pipe, dungarees, turtleneck sweater, and faded olive-drab army jacket, and said, "I never would have known. Glad to meet you." He extended a manicured hand, and Serpico remembered how soft it felt.

Ribustello asked Stanard how his Christmas had been, and how things in the division were.

"Everything's cool," Stanard replied.

"That's good. That's the way it should be."

Ribustello reached into his pocket and took out a roll of bills. He peeled off a couple of tens and handed them to Stanard. He peeled off two more for Serpico. "Here," he said lazily, "get yourself a hat."

A "hat" was a code word for a bonus above regularly scheduled payoffs. Serpico decided to play dumb. "I don't need any hats."

Ribustello's eyebrows arched slightly. "So buy yourself whatever you want," he said. "Get yourself something."

Serpico stared back at him, and then nodded at Stanard, and said, "Just give it to him," and walked out of the room. He went through the bar and out into the car.

A few minutes later, Stanard came outside, his face flushed. He got into the car. "Hey, Cheech, what did you do that for? You embarrassed the hell out of me."

After listening to Stanard's little talk about the 7th Division pad, Serpico had been determined to remain as diplomatic as possible in hopes of learning more about its operation, who was participating in it, and how high up it went. But he was so revolted by the spectacle of a police officer toadying to a racketeer like Ribustello—Ribustello offhandedly treating Stanard as if he were some sort of hired lackey, and Stanard accepting it—that he exploded with rage. "*I* embarrassed *you!*" he yelled. "Who the fuck told you to bring me here in the first place? I don't want any part of it. You were embarrassed, huh? How do you think I felt?"

Serpico was sure that this outburst would end any further relations with Stanard. But to his amazement, Stanard backed down. When he thought about it later, he realized that it was because Stanard had been so intent on bringing him into the fold, so certain that he had nothing to fear, that he felt he was above the law, immune to it.

"Well, I have to show you around," Stanard said. "We got a new man, we have to account for him, and I was just introducing you, that's all."

"Look," Serpico snapped, "I don't care what you do. You can do any fucking thing you want. I'm not interested. I have to make up my mind about things."

"OK, OK, take it easy. I can understand that." Stanard started the car. "Boy, you are one touchy guy." Stanard turned down 161st Street, a main thoroughfare running east and west through the South Bronx, and slowed as he approached the Bronx County Criminal Courthouse. A number of cops were congregated in front of it, and Stanard waved to them with easy familiarity. "Hey," he said, "there's the guy that's supposed to be your partner. He's been off." Stanard parked, and they walked back to the courthouse, and Serpico met Carmello (Gil) Zumatto, a burly, round-faced plainclothesman who looked like the actor Ernest Borgnine.

After a few minutes of chatting, Stanard said, "Listen, Gil, with Frank being new and all, if you don't want to work with him, I will."

"Nah," Zumatto said, "he was assigned to me. Besides, what the fuck, we're both *paisanos*. We'll get along OK."

Stanard glanced at his watch. "Jesus, look at the time. I'm starved. Let's grab a bite."

Serpico begged off, saying he had a dentist appointment. As he walked away from the courthouse, he looked quickly over his shoulder, and saw Stanard and Zumatto deep in conversation.

That night Serpico went home to his Greenwich Village apartment. He had spotted the ad for it and moved in just prior to his transfer to the 7th Division, and was still in the process of cleaning up the courtyard that had been billed as a garden.

The apartment at the time had only a bed, a table he had picked up somewhere, and a couple of chairs, and after the day he had spent with Stanard it was too depressing to be in. He did not feel like seeing anyone, so he took a long walk alone through the Village streets, ending up on one of the Hudson River piers, staring at the swirling black water, hunched against the wind. But the wind and the cold forced him off the pier, and he retreated to a dingy bar facing the river under the West Side Highway. There were two or three other customers, whom he guessed to be sailors. At the moment he wished he were shipping out somewhere with them. Serpico normally drank very little, but he had several Scotches in quick succession. They had no effect on him. Christ, he thought, I can't even get drunk.

He returned to the apartment and paced aimlessly back and forth. For the first time, he accepted the idea that he might have to leave the Police Department, that all his aspirations about being a cop, and all his concepts of what a cop stood for, were crashing around him. It was the waste, the waste of his hopes and the waste of the years he had put into the job, that angered him perhaps more than anything. He had wanted nothing to do with corruption. It had been thrust on him by crooked cops who seemed to have a free rein in the department. But if he did leave, it would not be without a fight. He would not be run out by the corruptors, and have to carry *that* humiliation with him for the rest of his life; first he would take them on, and the system that allowed them to operate.

Finally he phoned Captain Cornelius Behan again at home. It was the first time he had spoken to him since he had asked Behan to check into the 7th Division.

"Yes, Frank," Behan said, his voice, as usual, low and pensive, the kind Serpico recalled hearing as a boy in the darkness of the confessional.

"With all due respect, Captain," Serpico began brusquely, "what the hell have you got me into? This place which is supposed to be so clean is worse than the other place I was. The only difference is the price."

"I don't follow you, Frank."

"I'm telling you. They already told me I can get eight hundred a month for doing nothing, just like that."

"Oh, my God," Behan said, "what a bucket of worms."

"I don't know what to do, but I've got to do something. I'm not going to just sit here and take this."

"No, of course not, Frank. But are you sure of your facts?"

"Are you kidding?" Serpico shot back. "They put it right to me. They've got a hell of a pad going up there."

"A bucket of worms," Behan intoned again. "It sounds like a real bucket of worms."

"Captain, what should I do?"

"Frank, I don't know. I have to think about it."

Serpico continued to press him, and at last Behan said that perhaps they should meet to discuss it further. Since they were both taking night courses at the John Jay College of Criminal Justice, they could do it on an evening when their class schedules dovetailed. In the meantime, Behan said, he would attempt to put out a "feeler" about the matter to John Walsh, the First Deputy Commissioner. Serpico was impressed. Behan was obviously prepared to go much higher than his "clean as a hound's tooth" neighbor, Deputy Inspector Sheridan of the 7th Division. Walsh was the Police Department's number-two man—and in the view of most cops, he actually ran it. Serpico began to think that maybe something would come out of this after all.

A few nights later he saw Behan in a corridor of the college. "No, not here," Behan said, and they agreed to rendezvous on a nearby street corner. Serpico had his car, and Behan said that was perfect. He had to catch a train home from Pennsylvania Station, and they could talk while Serpico drove him there.

At the corner of Thirteenth Street and Fourth Avenue in Manhattan, Serpico watched as Behan approached, looking quite distinguished in his gray fedora and gray herringbone overcoat, carrying an attaché case. Serpico hoped he was as good as he looked. He was pinning everything on Behan, and he hardly knew him.

Behan got into the car, and Serpico pulled into a side street and double-parked. Almost mournfully, Behan asked once more if he was sure about what he had said on the telephone. Serpico related the two bar incidents he had personally witnessed with Stanard, and reaffirmed that one of the participants in the pad had told him that each man's share averaged eight hundred dollars a month. He did not, however, mention Stanard by name. He was afraid that if he did, Behan would immediately pass it on, and Stanard would be grabbed, and that would end the affair. This was the usual procedure in the department, but Serpico was bent on finding out how far and how high the 7th Division payoffs, and possibly others like them, went. Hanging a single cop was the easy way out, and he was not willing to settle for that.

It was the system that was corrupt, he said bitterly to Behan, and it had to be changed if honest cops were going to be able to operate effectively. Behan seemed to agree. At least he did not argue the point. Then he asked Serpico if he still wanted him to go to Commissioner Walsh, and when Serpico said yes, Behan replied that he had already been in contact with Walsh's office and expected to see Walsh himself fairly soon.

Although Behan did not spell it out to Serpico that night, a subsequent investigation showed that he had gone to a lieutenant in Walsh's office, a man who acted more or less as an administrative aide, and told him about a police officer in the 7th Division who had found evidence of corruption there and added on his own that the officer wanted to transfer out of plainclothes.

Under the circumstances, such a transfer would require Walsh's attention, and the lieutenant said he would see what he could do. Behan later acknowledged that at the time he thought Serpico had "exaggerated" the situation, that while Serpico had undoubtedly observed certain things which upset him, he was making "rather extensive" conclusions.

When Serpico dropped him off at Pennsylvania Station, Behan promised to keep in touch, and urged him not to give up hope, to have faith. He even suggested that Serpico come on one of the spiritual retreats he periodically organized. It would be good for his peace of mind, Behan said, warning him, however, that he would "get nothing more from it than what he put in."

Serpico nearly blurted out that it might be a bit awkward, since the inspector at the BCI who had accused him of being a homosexual regularly attended the retreats, but he managed to restrain himself. He thanked Behan, and said he would think it over.

When Serpico started working with Gil Zumatto, his assigned partner in the 7th Division, his imagination wasn't overly taxed as to what Zumatto and Stanard had discussed after he had left them in front of the Bronx County Criminal Courthouse.

Almost at once Zumatto asked him, "How do you feel about the money?"

Serpico repeated his litany of seeming indifference. "I don't care what you do, as long as I'm not involved," he said. "I don't want to get into any trouble."

Zumatto did not appear at all fazed by this. "Ah,

don't worry about it. I'll tell you what. I'll take your share and save it for you, and whenever you make up your mind, it'll be there."

Plainclothes life in the 7th Division was very relaxed. The men would sign in, check their mailboxes for any assignments or communications, then kill an hour or two in the coffee room, and perhaps take in a movie in the afternoon. A number of them lived in nearby suburban counties; some had swimming pools, and in the spring and summer especially, they would often while away their afternoons at one house or another, playing cards between dips.

They were supposed to ring in to the office regularly, but there was no switchboard control, nobody checked to see if this was being done, and it boiled down to a general rule to "keep in touch." So occasionally, either at poolside in the suburbs or playing pool in the back room of a bar, someone would say, "Hey, anybody call the office? Better give them a ring."

As Serpico soon observed, the main function of the division plainclothesmen was to protect the entire pad while servicing their racketeer clients. A principal safeguard was always to produce a minor arrest for the record whenever a complaint about illegal activity was passed down to the division for investigation. One day Zumatto said, "Come on, I got to check out some action."

A ghetto mother had reported wide-open gambling in her neighborhood and was afraid her teenage son was being sucked into becoming a policy runner, the lowest level in the numbers racket. Policy is one of the most lucrative underpinnings of organized crime. A runner dashes around from apartment to apartment and helps take bets for a collector in a particular area; next in the intricately structured racket is the pickup man, who brings the "work"—the betting slips—from various collectors to a controller. He in turn passes it on to a "banker," the money man. The spiral continues upward with many banks interlocked into still larger ones. Playing the numbers may be basically a "poor man's game," but it is still big business, and hundreds of millions of dollars are milked annually out of ghetto areas by the underworld.

When Serpico and Zumatto arrived on the block cited in the woman's letter, it was not long before they spotted the local collector. He was a "mover," going from place to place—an alley, a tenement, a candy store—to take his action. Zumatto watched him with some amusement until he finally said, "Let's grab him." They stopped the man, frisked him, and found enough slips and money on him to make an ironclad felony arrest. The collector was puzzled. "What's the matter?" he asked. "What's the problem? Ain't you from the division. You know, we're friends with the division. We're on."

The collector gave the name of the banker he worked for, and Zumatto said, "That's easy to check out. But we have a complaint from downtown, and we got to do something about it."

"Hey, that's cool, man. I understand, but I'm losing money just talking to you. I can't go in right now. This is prime time, you know. What say I meet you in front of the precinct at four-thirty? How's that?"

Zumatto smiled.

"I'll bring some work," the collector said. "I'll even bring my own work, and you won't have to worry about nothing."

"OK. Four-thirty, remember."

"No use breaking his chops," Zumatto said to Serpico afterward. "The guy he works for is good people. He's never late."

The incident had taken place in the 42nd Precinct, and later in the day Zumatto brought Serpico to a bar called the Piccadilly, across from the station house. Zumatto asked him if he wanted the collar. When Serpico said no, he didn't want this one, Zumatto looked around the bar, spied another plainclothesman, and offered it to him. He was delighted to take it, and they all adjourned to the sidewalk in front of the station house. Promptly at four-thirty, the collector, whose name Serpico learned was Brook Sims, walked up, smiling, clutching a handful of slips, although only enough for a misdemeanor arrest. The third plainclothesman marched Sims up to the desk and booked him. The case was dismissed the next day in court, but if anyone checked the record, the complaint had been investigated and an arrest had been made.

Somebody in the policy racket who did not pay off was treated a good deal less kindly. Serpico was driving down a street with his partner when Zumatto suddenly said, "Stop, I want to talk to that guy." The object of Zumatto's attention was a huge black man named Lloyd Hassel. Zumatto hurried after him, and Serpico followed, thinking he might need help. But Hassel just stood there, a sheepish grin on his face.

"You got it yet?" Zumatto asked.

"No, man, no. Look, I'm really strung out. I'm trying real hard to get it together. I'll have it for you sure tonight."

The next day Zumatto said to Serpico, "He didn't show up. Let's find him." Zumatto cruised the neighborhood. Repeatedly he got out of the car to question someone on the street, in several bars, and two or three hallways. He could not have been more energetic, Serpico thought, if he had been working on the crime of the century.

"I don't get it," Serpico said. "He's just a collector. Isn't his boss taking care of him?"

"That's all he is," Zumatto said, "a half-assed collector. He used to be one of the biggest operators around, but he got wiped out. I don't know how. Who knows with these fucks? He's trying to put it together again. He's working for some guy, but the guy isn't paying for him, so he's got to pay for his location."

Finally Hassel was spotted in a tenement doorway. Zumatto bolted out of the car and pushed him into the hall, and came up with a handful of policy slips. "Now, you fuck, you're going."

"Man, give me a break," Hassel pleaded, his huge bulk cowering in the hallway.

"I told you to get the fucking money up, and you keep jerking me around, telling me you'll be there, and you're never there."

Serpico saw an elderly black woman watch for a moment through a slightly opened door behind Hassel. Then she quietly closed it.

"I'm on the balls of my ass, man," Hassel said. "I'm

trying real hard. Tonight, for sure. I'll have it. I'll have it."

"This is the last fucking time," Zumatto said. "If you don't get it up this time, you're going."

In the morning Zumatto greeted Serpico with a smile. "Well," he said, "the fuck showed up. I mean, if you let them get away with that shit, think of what it could start."

The services provided by the division plainclothesmen to their clients were matched only by their greed, and Serpico never ceased to be amazed at the way Stanard continued to flaunt both. During a day when Zumatto was in court, Serpico worked with Stanard. They pulled up in front of a house in a quiet residential Puerto Rican area, and Stanard waved to a woman on the second-story porch. A few minutes later a man appeared at the door, and walked over to the car. "This is Frank," Stanard said. "He's all right, he's one of the boys." The man, a Puerto Rican, shook Serpico's hand, and then Stanard said, "The reason I came down is the precinct is taking warrants out on a couple of your locations. The candy store and the tailor shop."

The man thanked Stanard. He said he would close his betting operation in the two stores until things calmed down, and went back into the house.

Stanard drove angrily away from the curb, tires squealing. "That cheap bastard," he said. "Now, if it was a wop, he would've been good for something. Maybe even a fifty or at least a good bottle of Scotch."

Not long afterward Serpico and Zumatto stopped at a luncheonette that was a favorite division hangout for breakfast. A couple of other plainclothesmen were in a rear booth, and Serpico saw several boxes of .25-caliber ammunition open on the table next to one of them. A stocky, coarse-faced man, well dressed except that he was not wearing a tie, walked in and sat in the booth. Serpico watched as the stocky man took the ammunition and even showed off the gun it was meant for.

When Serpico and Zumatto left the luncheonette, Serpico asked, "Who's the guy who got the ammo?"

"Oh, that's Vic," Zumatto said airily, "Vic Gutierrez. He's a good informant."

Victor Gutierrez was actually a small-time racketeer, and one of his best rackets was letting the 7th Division plainclothesmen know when a new policy operation started up within the division boundaries. If the operation was big enough, it would be forced on the pad, and Gutierrez would be rewarded with a piece of the action.

What Serpico found so ironic was that Stanard, Zumatto, and the other plainclothesmen he met were really professional in the sense that they were first-class investigators, and they brought to their craft all the requisites this entailed—instinct, patience, technique, determination, and accurate intelligence provided by a carefully nurtured network of informers. If they had wanted to, they could have wiped out a major portion of their number-one target—illegal gambling—practically overnight. But their motivation instead was that there was money in it for them. Serpico was constantly impressed by the way they could ferret out operations no matter how cleverly concealed, but their purpose was always to extort money from the people they caught.

When an East Harlem numbers banker expanded into the South Bronx without going on the pad, Zumatto and two other plainclothesmen, aided by a tip from Gutierrez, soon pinpointed one of his "drops"—a collection place for betting slips and money—and raided it. When the money on the table was counted, it amounted to four thousand dollars. The plainclothesmen demanded more, and one of the banker's controllers called his boss. He arrived with another thousand, but it still wasn't enough, and Zumatto accompanied the controller back to East Harlem for an additional two thousand.

In an individual instance of graft like this—or "score," as it was called—the cops involved would keep the money. Stanard, for example, once boasted to Serpico that he had made "sixty big ones"—sixty thousand dollars, tax-free, of course—in the previous two years alone as a result of his scores and his share of the pad.

Serpico heard all the rationalizations. They ranged from the cherished tenet that people were going to gamble anyway, to one that held that a cop had a thankless job and was reaping a well-deserved reward by confiscating the money instead of turning it over to the city, where it would wind up in the hands of welfare-chiselers. Many people might agree with Stanard that it was just gambling money and therefore "clean." These same people would throw up their hands in horror at the idea of similar payoffs in heroin traffic, although Serpico later found out that this philosophy of clean money was very flexible indeed.

There was also the general breakdown this corruption wrought in police effectiveness. There wasn't a kid in the ghettos who did not know what was going on, and yet whenever Serpico picked up a newspaper or tuned in his television set, some politician was yammering about the decline of law and order. As a drawing-room abstraction, this paradox might be obvious. But it was one Frank Serpico had to face personally every time he went into the street.

chapter 10

About three weeks after their initial meeting, Captain Behan reported to Serpico what seemed to be good news. He said that he had been summoned to Commissioner Walsh's office, and had told him about the conditions Serpico had discovered in the 7th Division. Walsh was delighted that a "man of integrity had surfaced," and that he could count on an "honest officer in the field who was willing to talk about corruption."

Behan added that he had discussed with Walsh the possibility of secretly transferring Serpico out of the division into an anticorruption squad where he could follow up what he had learned on his own, but Walsh had said that he did not think it "advisable" to do this at the present time. He "preferred" to leave him where he was, in the 7th Division, to pass on more information.

This second part of Behan's message left Serpico very uneasy. He did not cotton at all to the idea of being put in a kind of limbo, lacking any official investigative status, without any direct contact with Walsh. And he was worried about how long he could maintain the delicate position he was in without arousing the suspicions of the other 7th Division plainclothesmen.

When Behan said that Walsh wanted them to meet again so Serpico could provide additional details about what was going on in the division, Serpico replied that he would call back to set a date, but he did not. His disappointment and concern mounted at not having been immediately assigned to an anticorruption unit, and he desperately needed someone to confide in.

Ever since Captain Foran's warning that he risked being found "facedown in the East River" if he pursued the issue of police corruption, Serpico had avoided David Durk. But Durk, who was still working under Foran at the Department of Investigation, kept phoning him. He felt "nebulous," "disappointed," and "embarrassed" about Foran's conduct, he said to Serpico, but on the other hand Foran probably "was expressing the reality of the situation" in the Police Department. Durk insisted that he and Serpico had to get together, that they had a mutual interest in cleaning things up which was too important to throw away. Finally Serpico agreed to see him, and subsequently told him about what he was encountering in the 7th

Division, about how he had gone to Captain Behan, and how Behan had informed Commissioner Walsh. Despite Serpico's qualms, Durk appeared impressed by the news.

In the middle of February 1967 Behan contacted Serpico once more about the meeting, and it was fixed for Sunday afternoon, February 19, on a ramp over the Van Wyck Expressway in Queens, on the way to Kennedy Airport. Serpico spoke to Durk about it, and Durk asked to go along, arguing that if something went wrong he would be a witness to the fact that the meeting had taken place.

Serpico decided to let him come. When they got there, Behan was parked by the side of the road. It was a dark, gray day, and Serpico walked through a light drizzle to Behan's car while Durk remained behind. As soon as Serpico got in Behan asked, "Who's that with you?"

"He's a friend of mine, a police officer. He drove me here. Can he come over?"

Behan became visibly upset. "My God, no, this is all supposed to be confidential. You know that."

Serpico blew up. "What the hell am I supposed to do?" he snapped. "I've got to talk to somebody."

Perhaps he understood the obvious strain Serpico was under; at any rate, Behan dropped the subject. After a moment he went on. "As I told you, I've notified Commissioner Walsh of everything you've told me so far, and what he'd really like you to do is stay there and advise him of everything that goes on. What's your decision?"

Serpico hesitated. He knew what he wanted, but he

was in an extremely awkward spot. John F. Walsh was a fearsome name in the Police Department. Even his closest associates were uncomfortable in the presence of this tight-lipped, cold-eyed man. His star had risen swiftly in the great scandal that hit the department in the early 1950s when the bookmaker Harry Gross seemed to have almost as many cops on the payroll as the city did, and as the head of his own confidential squad he had built a reputation as a relentless disciplinarian. In one of his more famous remarks, which he chose to deliver at a headquarters Christmas party for over two hundred ranking officers, he said, "I am out to catch you if you do anything to besmirch this department." It was an interesting exercise in internal image-making, for although Walsh's investigators—or "shoo-flies"—caught countless cops in minor violations of the department's rules and procedures, they somehow turned up very little graft.

Nonetheless, for Serpico as for most cops, Walsh *was* the Police Department. He ran it. Police Commissioners would come and go, but First Deputy Commissioner Walsh remained. He had served in that post under four of them, gradually increasing his power. The First Deputy assumed command of the department whenever the Police Commissioner was out of town, and since Mayor Lindsay's appointee, Howard Leary, was in the process of setting a modern record for absenteeism, Walsh had entrenched himself more firmly than ever. With Leary's blessing, Walsh had gathered all of the department's various anticorruption units under his control, had stripped away much of the authority of the Chief Inspector, nominally in tactical command of the department's manpower, and had begun to nibble successfully at many of the prerogatives of the Chief of Detectives, including the promotion of plainclothesmen to detective rank.

Serpico at last said to Behan, "I'll do whatever Commissioner Walsh feels would be most productive."

"Well," Behan said with evident relief, "he wants you to stay where you are. That's what he basically wants you to do."

Serpico tried to explain how precarious he felt his situation was in the 7th Division. What could be done about that? Behan promptly replied, "Frank, I'm glad you asked. The Commissioner says he'll meet you at any time of the day or night at any place you say. He'll meet you at the last station of any subway in the city, if necessary. That's how determined he is to protect your identity."

"OK, fine," Serpico said, "that sounds good."

Behan said that Walsh had wanted additional details about the corruption in the division, and it was then that Serpico gave him the names, told him that Stanard seemed to be running the pad and that Zumatto was one of his key aides.

"The thing for you now is to go back there and keep your eyes open," Behan concluded. "The Commissioner will be in touch with you."

Serpico rejoined Durk. "Well?" Durk asked.

"He says I'll be working directly for Walsh," Serpico said. "On a confidential basis."

Durk continued to read hopeful signs in all of this, but Serpico was still gripped with all kinds of doubt.

He was the one on the spot, after all, and there were too many loose ends. How was he to get in touch with Walsh, or Walsh with him? What kind of investigative procedure should he follow, how deeply should he allow himself to get involved in the 7th Division pad? He had not asked Behan because he sensed that Behan himself did not have the answers, that he probably had not dared to raise them with Walsh.

On February 25 he spoke to Behan again on the phone. Behan said that he had seen Walsh and had relayed what transpired between them, and they arranged another meeting for March 2. It never came off. On March 1, Behan was hospitalized for most of a month with a leg infection. Serpico didn't know this at the time, but he was not disturbed about the lack of word from Behan. The man he was waiting to hear from was John Walsh.

Serpico's major worry was what he would do once he was eligible for a share of the 7th Division pad, but Gil Zumatto's offer to hold his portion of the money until he decided whether or not he wanted it gave him some breathing room.

Zumatto was able to make such a proposal because he was one of the three "bagmen" in the division who picked up the payoffs, usually on a twice-a-month basis, and sat in when the money was split up. It was not long before Zumatto took Serpico along with him on some of his stops. Once Zumatto turned to him and said, "Hey, I hope you don't embarrass me, because I told the guys you were all right." Serpico could not tell if Zumatto was apprehensive about being reported for graft, or if he had guessed that Serpico would never claim the money due him and was concerned that the other men in the division might find out that Zumatto was getting a double share. The latter turned out to be the case.

Zumatto kept an extra apartment in the Bronx for, as he said, "some socializing," and occasionally he would take Serpico there while he counted a day's take. A tall wooden cabinet with glass doors stood against one wall, and on the top shelf there were a number of envelopes. Zumatto would point to one of them and say, "Well, whenever you want it, Frank, it's right up here. Just let me know, and you got it, down to the penny. Like I said, all I'm doing is keeping it for you. But if it's OK, I'll take ten bucks out for the clerical men, so we don't get those bullshit complaints and they won't break our chops on the paperwork."

Perhaps to further the illusion that Serpico was taking the money, Zumatto brought him along when he attended a division meeting about the pad. These meetings, with Stanard acting as chairman, were often held in motels, although the one Zumatto took him to was in the back room of a saloon not far from 7th Division headquarters. Serpico waited at the bar in front while it went on, but using the excuse of telling Zumatto that he had to go somewhere, he did step into the room once to observe the scene, and found a half dozen or so plainclothesmen around a table drinking and slicing chunks of meat from a large ham, and Stanard frowning and saying about one of the gamblers on the pad, "No, I think he's getting a little hot, and we're going to have to lay him off for a month and see what happens." And again Serpico thought that if they only applied half the care and ingenuity they put into the pad to law enforcement, the effect on the city would be magical.

Then, toward the middle of April, Serpico's position in the 7th Division became far more hazardous. "Frank," Zumatto said, "I think I'm going to be transferred."

Serpico had grown increasingly edgy at the absence of any sign that Walsh even knew he existed, and now with Zumatto indicating that he expected a transfer, it was only a matter of time before the fiction that Serpico had been taking his share of the pad would be discovered.

He telephoned Behan and said that he had to see him as soon as possible. They met on the evening of April 12 at Thirteenth Street and Seventh Avenue, on the border of Greenwich Village, near Serpico's apartment.

Why, Serpico demanded, had Walsh not contacted him yet, or given Behan a number where he could be reached directly? Behan became visibly unstrung. He was, Serpico suddenly realized, terrified of Walsh. Serpico felt as though he had been inquiring about God when Behan finally replied that the First Deputy Commissioner "often worked in strange ways." Serpico then described the spot he would be in if Zumatto left. He told Behan that the pad continued to flourish; that pickups occurred twice a month; that there was at least one meeting a month to decide who was on or off the pad; that policy numbers dominated the pad because bookmaking relied heavily on the telephone and there was always the possibility of wiretaps; and that individual scores were permitted so long as the person being scored was not a "cousin," not someone already on the pad. Serpico explained that he had not probed deeper—had not, for example, gone to the monthly meetings—because he thought this deep an involvement required official sanction and direction.

Behan assured Serpico that he planned to confer with Walsh the next day, and he subsequently reported that he had passed on what Serpico had said. Commissioner Walsh, he added, was extremely appreciative and would be "reaching out" to Serpico. "When," Serpico wanted to know, not bothering to mask his sarcasm, "will that be?"

Behan's voice, usually so measured, took on a sharp tone of annoyance. He had done all he could, he said, and the whole affair was now completely out of his hands. "In no way am I to be considered an intermediary any longer." A tremulous rush of words followed. He was not in a position to do more, he said. He had not been working in an official capacity in the first place as far as Serpico's problems were concerned, but had merely tried to help in the best way he knew how, had indeed told Walsh everything Serpico had told him, and it was up to Walsh to do the rest. Clearly Behan wished that he had never heard of Frank Serpico.

Serpico was stunned by Behan's abrupt self-removal from the picture. Despite his impatience at not having been contacted by Walsh, he had consoled himself with the idea that he was, after all, just a cop, and that Behan was a captain acting on his behalf with the First Deputy Commissioner. But he had assumed that they knew what they were doing, that eventually he would receive instructions at home some night, or for that matter every night, as to his future moves in the 7th Division.

Now that expectation seemed to have vanished. He felt that he was in the middle of a monstrous runaround. But his initial shock gave way to anger. He still was not going to give up.

All along, he had been seeing David Durk, perhaps once a week. Durk would call him and come down to his apartment and have a couple of drinks and ramble on about how it would be if they could ever work together. Imagine what it would be like, Durk would exclaim, if they were given a radio car and turned loose on the city! However fanciful this prospect was, Serpico willingly went along with it; if nothing else, it was a welcome relief from the reality of the South Bronx.

To Serpico's surprise, Durk did not share his pessimism when he spoke periodically of the fact that he had heard nothing from Walsh. Things would work out, Durk said enigmatically, "changes are going to be made." He maintained this attitude even after Serpico told him what Behan had said. And then Serpico found out why. Durk had been to his friend at City Hall, Jay Kriegel, and had given him a general rundown of the corruption Serpico had encountered. Kriegel really cared about police corruption, Durk insisted, and what's more, could do something about it. Durk said he had rounded up some other cops who were unhappy with conditions in the department, and through Kriegel they would all meet with Mayor Lindsay himself and launch a reformation once and for all. The thing now was for Serpico to let Durk take him to Kriegel, so that Kriegel could listen firsthand as Serpico spelled out what was going on in the 7th Division.

Serpico liked the idea. He had come more and more to the conclusion that the Police Department was incapable of acting on itself from within. Superior officers who spent twenty-odd years coming up through the ranks had seen all the corruption on the way, had done nothing about it, and had necessarily developed alliances and taken positions that precluded the possibility of changing the *status quo*. If changes did occur, they would have to be triggered from the outside, and Jay Kriegel appeared to be an ideal catalyst. There would be nothing awkward about Kriegel's getting involved in police matters; one of his primary responsibilities as a member of the Mayor's staff was the Police Department.

Jay Kriegel was not yet twenty-seven years old, had grown up in Brooklyn, and after Amherst and Harvard Law School had burst practically full-blown onto the Lindsay team. Elfin-bodied, bespectacled, with a mop of frizzy hair, and slavishly devoted to the Mayor, he ordinarily worked twelve to fifteen hours a day, seven days a week, as John Lindsay's ace troubleshooter, spewing forth half-finished sentences at machine-gun speed, his mind always whirling, simultaneously handling projects and problems for the administration in the city, at the state capitol in Albany, in Washington, D.C. He relished the frenetic pace, he told a reporter; when everything was going well, it meant he had "thirty or forty things in the air," knew precisely where each one was, and could reach out and grab it whenever he had to.

Serpico was ushered into Kriegel's cluttered basement office at City Hall on a Sunday afternoon in late April. It was the first time he had seen Kriegel and he remembered being amazed that such a youthful little man was such an important figure in the Lindsay administration. He looked, Serpico thought, just like the comedian Woody Allen.

Under Durk's prompting, Serpico began a two-hour recital of the rumors that he had heard about plainclothes, then all the circumstances surrounding the envelope with three hundred dollars. Kriegel listened to him raptly, sometimes scribbling a note on a legal-size pad of paper, occasionally shaking his head and interjecting a "My God, this is unbelievable" or "Wow!"

Serpico went through his entire experience in the 7th Division. Kriegel peppered him with questions. How big was the "nut"—each plainclothesman's share of the pad? Who was involved? How high did it go? Once he remarked that he had heard that things like this were going on, but never in this way, never directly from a cop laying it on the line as Serpico was.

Serpico said that as far as he knew every plainclothesman in the division except himself was in on the pad, and one lieutenant, possibly two. He added that he felt reasonably sure that the other men at the borough and precinct level were raking in graft as well. That was precisely the point. By himself, without support, it was difficult to say exactly how widespread the corruption was, and where it would lead. The only way to determine its extent was through a full-scale investigation, Serpico said, utilizing all the resources the department theoretically had at its disposal—cameras, electronic bugs, surveillance, undercover men.

And he recounted to Kriegel, emphasizing his frustration, how he had contacted Behan, how Behan had gone to Walsh, and how absolutely nothing had come of it.

Kriegel mumbled something Serpico could not catch and shook his head again. Durk broke in to say that Serpico was in a dangerous predicament in the 7th Division, and that if the situation were allowed to continue, the least that should be done was to get him out of the South Bronx.

Kriegel responded at once that this was something he had already recognized, and he would get to work on it, although he did not specify what he had in mind. He would talk to the Mayor, he said, and arrive at a course of action, and he thanked Serpico, expressing his sympathy for what he had been going through.

Several days later Durk came to Serpico's apartment. Serpico poured him a drink. "What's with your friend Kriegel?"

Durk began raging. "That fuck Jay," Durk said, "he's not going to do a fucking thing either." According to

Durk, Kriegel told him that he had spoken to the Mayor—"Big John," as Durk called him—and the meeting Durk had proposed with Lindsay had been called off. "Big John" could not afford upsetting the police, because of the prospect of another "long, hot summer" in the ghettos. Whether this quote was the Mayor's or Kriegel's, or simply Durk's interpretation of what was said, Frank Serpico would never know, nor did he care.

All he knew was that he had reached another dead end.

chapter 11

Serpico's partner, Gil Zumatto, was transferred on May 25, and Serpico was teamed with another veteran member of the 7th Division.

The vacancy caused by Zumatto's departure was also filled the same day by a dapper, sleekly handsome plainclothesman named James Paretti. He was fresh from the Chief Inspector's office, and the word around the division was to "watch this guy." Serpico, ready to grasp any straw, wondered if he might be an undercover man dispatched by Walsh, and he waited hopefully for some indication from Paretti that this was the case.

David Durk, meanwhile, had been back to Serpico's apartment several times. He would sit in the rock garden Serpico had fashioned and, drink in hand, continue to rant about Kriegel's deviousness. "Cut out the bullshit," Serpico finally told him. He was fed up with hearing this over and over again. Durk became defensive, and a violent argument followed. Didn't Serpico trust him? Durk demanded. Was he implying that Durk was responsible for the fact that nothing had come out of the sessions with Captain Foran and Kriegel?

"No, I never said that," Serpico replied. "But I'm tired of you building my hopes all the time with these bullshit big names of yours, and then finding out that nobody wants to get involved."

"Well, if you don't trust me, I'll walk out right now, and that'll be it. Just say the word."

"For Christ's sake, Durk," Serpico said, "pour yourself another drink."

Then the night Zumatto was transferred Durk called excitedly to say there was nothing more to worry about. At long last he had come up with the answer to everything. He had arranged an appointment on May 30, Memorial Day, for both of them to see Mayor Lindsay's Commissioner of Investigation, Arnold Fraiman, at Fraiman's Manhattan apartment. Serpico asked what good would that do? Wasn't Captain Foran still working for Fraiman, as Durk was, in the Department of Investigation? They could bypass Foran, Durk explained. He had finally developed a close rapport with Fraiman, something that he did not enjoy at the time they had gone to Captain Foran with the three-hundred-dollar envelope. Durk said that he had had several long discussions with Fraiman in which he had pointed out that while the Department of Investigation had a mandate to look into any municipal agency, in practice an allegation involving corrupt cops was turned over to the Police Department itself to investigate. Fraiman, according to Durk, was enthusiastic about the possibility of developing his own police-corruption cases. Durk said he told Fraiman that he knew just the police officer who could help, and Fraiman was anxious to meet him.

"Sure, Durk," Serpico said.

"Frank, I'm telling you, he wants to do a job. He isn't a political fuck like Kriegel. You give him something, he moves on it."

"Sure, Durk."

"Frank, listen to me, I'm telling you this is the way to do it."

In the end, Serpico figured he had little to lose, and he agreed to see Fraiman. Then, late on Memorial Day afternoon on the way to Fraiman's apartment, he got his first inkling that there might be some political animal in Fraiman after all. Durk told him not to mention having been to Kriegel's office. There was, as Durk put it, no point in complicating things.

There was also no confrontation more predictably ill-fated than the one between Serpico and Fraiman. Arnold Guy Fraiman, square-jawed and athletic, his voice haughty, eyes disdainful, his graying hair styled in a boyish crewcut, was a man who carefully cultivated a patrician air. One visualized him stepping off a tennis court after several sets without a bead of sweat on him, or leaning back comfortably in the deep leather chair of an exclusive men's club, ready to savor cigars and brandy. Even his warmest admirers admitted that Arnold Fraiman set great store by appearances, his own as well as others', and modesty was not counted among his virtues.

Serpico sensed this at once. Fraiman greeted him and Durk at the door in pressed gray flannel slacks, polished slip-ons, a white shirt open at the neck, a cashmere cardigan sweater. His eyes flickered over Serpico, over his shock of hair and bushy beard, the scuffed boots, faded corduroy trousers, and black turtleneck sweater, over the gold Winnie-the-Pooh dangling from his neck. Serpico could almost hear Fraiman thinking, *What does this guy want, what can he tell me?*

"Commissioner," Durk said, "this is Frank Serpico, the man I've been telling you about. Now you'll hear what he has to say with your own ears." Fraiman extended a hand—somewhat gingerly, Serpico thought. "Nice to see you," he said.

They gathered around a coffee table in the living room, Fraiman and Durk with Scotches, Serpico settling for a glass of plain soda water.

After an inquiring glance from Fraiman, and an introduction from Durk attesting to Serpico's dedication as a cop, he told his story again. He decided to make an effort to try to break through the hauteur he felt in Fraiman, and whenever he could, he worked in a respectful "sir." It was a disconcerting experience. Fraiman asked no questions. Occasionally he would reach for a nut from a bowl on the table, and once or twice he looked at Serpico, but mostly he stared into space.

When Serpico had finished, there was a long pause. "Well," he heard Fraiman say, "what do you want me to do about it?"

Serpico had gone to the meeting with a good deal of skepticism about whether anything would come out of it, but he was flabbergasted by Fraiman's question, and he struggled to control his anger. He wanted to shout out, "What the hell are you asking *me* for? I'm just a dumb cop, right? *You're* the big Commissioner of Investigation, and *you're* asking *me* what you should do?"

He resisted the temptation, however, and replied, "Well, sir, there are a lot of things you can do. You can bug the division office, for openers."

Fraiman was already shaking his head. That would require a court order, he said. It would mean having to go outside his department, too many people would inevitably learn what was afoot. Then he asked Serpico if he would wear a recorder.

Serpico replied that this would not be very productive. At best, all he could get would be other policemen at his level, when the important thing was to discover how high the corruption went. Besides, he continued, with Zumatto leaving, the rest of the plainclothesmen were on the verge of discovering that he had not been sharing in the pad, and it was unlikely any of them would discuss payoffs with him.

Serpico had not understood Fraiman's reference to a court order for the bug he had suggested, and he belatedly realized that Fraiman must be confusing a bug with a telephone wiretap. He explained that by a bug he meant an electronic device that transmitted signals which could be picked up by someone stationed a couple of blocks away. This, Serpico found himself telling the Commissioner of Investigation, did not need court authorization at the time. And if Fraiman was worried about going into the division office, Serpico had another idea. The 7th Division plainclothesmen often used a panel truck with peepholes in it for surveillance work. Serpico had been in it himself, and he knew how the men, bored with the duty, were liable to talk about anything. The truck, he said, was being brought into a Police Department garage in Manhattan for servicing, and it would be a simple matter for him and Durk to get in there to install the bug.

Fraiman was still hesitant, and Durk broke in to urge his approval. There was more talk about placing the bug in the truck, and finally Fraiman agreed that it was the thing to do. Then at the door, as they were leaving, Serpico noticed a doubtful look on Fraiman's face again. "Yes, well, OK," Fraiman said, as if he was wondering whether or not he had made the right decision.

After they had left Durk brushed aside Serpico's misgivings. "Come on, Frank," he said, "cut out the goddamn paranoia. It's all set."

Two nights later Durk rang the bell at Serpico's apartment and asked him to come out to his car. In the car, Durk reached into the glove compartment and took out a small object. He held it in the palm of his hand briefly before putting it back. "Well, I've got it," he said.

Serpico was elated. Now, he thought, a responsible agency was actually going to hear what he had been trying to tell so many people for so long. And he had another plan. The surveillance truck was not due for servicing for another day or so, and instead of risking the installation of the bug in the garage, he would drive it down from the Bronx himself, and they could put the bug in before the truck ever got to the garage.

Along with his elation Serpico felt very badly about having been so hard on Durk, and for having completely misjudged Fraiman. It was a good lesson, he told himself, to remember in the future.

Durk said that he would be in touch the next day to find out when Serpico would be bringing the truck down. But when he called, there was a familiar string of curses. He had been ordered to return the device to the Department of Investigation because it was needed in a bribery case involving a municipal building inspector. Serpico was somewhat irritated by the news, but it was not the end of the world. When, he asked, would they be able to have it?

Then Durk had to admit the truth—the whole project was off. Commissioner Fraiman had decided, Durk said, that Serpico was a "psycho."

It was the first week of June, collection time for the 7th Division plainclothes pad. In the car Serpico's new partner studied the list of stops he had to make. As Serpico knew, the list, for security reasons, contained only the names or code names of the numbers bankers and controllers who were paying off, and not their addresses. "Hell," the plainclothesman said, "I don't know where half these places are exactly," and turned hopefully to Serpico.

"Well, you know, I never really bothered to notice," Serpico said.

His partner grunted, and said, "This is going to take all goddamn day." When he did not know where a particular payoff stop was, he would have to walk around the neighborhood until he found a street collector who could give him directions. The operations were usually so open that this was no problem, except that it required time.

Serpico rather liked this new man he had been teamed up with. Thin, bespectacled, olive-skinned, with a slight stoop, his hair thinning back from a widow's peak, his manner soft-spoken, he looked like someone who might be found behind a druggist's counter. As they drove around, he began to wax philosophical about the pad. "I don't know what the hell I'm doing this for," he said, and reflected morosely on how he could ever face his family if something went wrong and what he was doing got out.

"Why take a chance?" Serpico asked.

He shrugged his shoulders. "Hey, Frank, everybody does it. You know that." He brightened a bit at the thought, and said that the pad was one of the fringe benefits in being a plainclothesman. The big thing, he said, was not to get too greedy; if you didn't get too greedy, there would be no trouble.

He lapsed into silence, and his spirits seemed to sag again. Besides, he said, the whole thing was expected of you, this was what you had to do, and you just went along with it. He had once tried to stop sharing in the

pad, he implied, but he grew vague at this point, and
Serpico could not tell whether he had been talked out
of it, or simply turned down. The memory of that
time, however, apparently fired some bitter embers inside the man, and he suddenly declared that maybe
he would get out of the pad someday, that if necessary
he would go back to the "bag"—police slang for uniformed duty.

Serpico half believed him. This, he thought, was
how the system worked. There were thousands of
cops who wanted to be honest, but the system did not
let them. Somehow everything had got twisted so that
the honest cop had to fear the crooked ones, instead
of the other way around.

After each stop, the wad of bills Serpico's partner
carried would swell. Finally it grew so large that he
was afraid to keep it on him, and he reached over and
put the money in the glove compartment, telling Serpico that he would be right back, that he had to go
around the corner for a pickup.

"Don't leave that with me," Serpico said. "What if
I get grabbed?"

"Nobody's going to grab you here," the plainclothesman insisted. "The place is a block away. I'm
the one that's got to worry. Look, I'll stuff it under the
seat." Then he walked off, while Serpico spent an uncomfortable ten minutes imagining being caught
alone in the car with the money.

At the end of the day Serpico's partner had so much
cash on him that he had to separate it into two rolls.
As he sat in the car doing this, he began peeling off

hundred-dollar bills and said to Serpico, "Here. I might as well give it to you now."

Serpico had wondered under what circumstances the realization would come that he had not been participating in the pad, and how he would handle it, and now he simply stared blankly back and said, "What's that?"

"That's yours."

"What do you mean?"

"Hey, is this a gag, Frank? What do you mean, what do I mean? That's yours for the month." He began to look somewhat disturbed. He peered curiously at Serpico.

"But I haven't been taking any money," Serpico said. He was looking directly at his partner as he spoke, and he saw the color in his face literally vanish.

The man started to tremble. "Frank, what are you trying to do to me?"

"I'm not trying to do anything."

"But what were you doing with Zumatto?"

"Why don't you ask him?"

"I mean, what about your nut? Where's it been going?"

"I told you," Serpico said quietly. "Ask Zumatto. He claims he's been stashing it for me."

As they drove back to the division office his partner was in a trance, his face still ashen, and Serpico thought he was on the verge of a coronary seizure. The news that Serpico had not been participating in the pad spread quickly through the plainclothes ranks, and a meeting for the next day was hurriedly called by Stanard.

"You *be* there," he told Serpico.

Serpico was dumbfounded by the site Stanard picked for the meeting. Instead of being in the back room of a bar, or in an obscure motel, as he would have supposed, it was held across the street from a small, triangular, tree-shaded park in front of the Bronx County Criminal Courthouse. To Serpico this was the most awesome example yet of the total arrogance of police graft. At midday when everyone gathered around Stanard there wasn't a busier spot in the South Bronx. There they were, under a tree—about half the division plainclothesmen, including a lieutenant—openly discussing the pad while a parade of judges, cops, lawyers, assistant district attorneys, reporters, crooks, and ordinary citizens walked by.

Even more amazing was the meeting itself. Serpico had fully expected, braced himself, to be the target of accusations that he was a spy sent in by a headquarters investigating unit. But the whole thrust of what was said, all the fury and suspicion of the men, was directed at Zumatto, for taking a double share. "That conniving bastard," one of them said, "I'll cut his fucking balls off. That's my money he was stealing."

Stanard tried to calm everybody down. "Don't worry about Zumatto," he said. "I'll straighten him out. I'll get the money back. And if I get any flak from him, he doesn't get his severance." The reason he had brought them together, he said, was to tighten up discipline regarding the pad and to make sure that what Zumatto had done was not repeated. Instead of only three bagmen picking up the payoffs for everyone else, he proposed that all the plainclothesmen would have to make collections. "If you don't make any stops," Stanard said, "you don't get any bread."

After some give-and-take, the men voted in favor of this, and Stanard turned to Serpico. "How do you feel about it, Frank?"

Serpico was not going to become involved in the pad, but he did not want to alienate the men either. "No," he said. "I haven't taken any money, and it's not worth my while to start now."

Stanard tried to argue with him. "Well, if you start now, we could make it up to you. We'll get you that money Zumatto took, and, you know, you got a lot of years left."

"Look, I'd rather not. Let it go the way it's been going, and that's that."

"I'll tell you what. We'll give you a hundred a month for your expenses."

"Why should I take a hundred? If I'm going to take a hundred bucks, I might as well take the whole thing."

"Yeah, but Frank, it's just expense money. You got to lay it out to the desk officers anyway." Stanard was talking about the cash that plainclothesmen had to pay the clerical people whenever they made an arrest.

"What do you mean, expense money?" Serpico asked, "It's coming from the same place, isn't it?"

"OK, Frank, what do you want to do?"

"I told you. I just want to be left alone and do my own thing."

"Well, OK, this is how it's going to be then,"

Stanard said, addressing the entire group, an air of finality in his voice. "Whoever is working with Frank pays for his lunch and everything like that, and at the end of the day you submit it, and you'll be taken care of."

Serpico decided he had gone about as far as he could without stirring up a whole new round of suspicions, and he did not protest. He knew Stanard had been trying to suck him in any way he could. Whether it was called "expense" money or whatever, if he accepted so much as a dollar, he would be as tainted as the others. At least with meals, no cash would change hands. If his partner bought his lunch or dinner, he would simply pick up the tab the next day. If he knew anything about these guys, he thought, it was their attitude about money. Not one of them would raise a finger if he offered to buy a meal, and would undoubtedly put in an expense chit for it to boot.

One of the plainclothesmen in the division fell in step with him as they walked away from the meeting and suggested stopping in a bar for a beer.

After they had been served, the plainclothesman said, "Hey, I don't get it."

"What?"

"Frank, why don't you take the money?"

"I don't need the fucking money," Serpico said. "What'll I do with it?"

"Jeez, I don't know, whatever you want." The plainclothesman studied his beer, considering the problem. Suddenly he brightened. "You could give it away. Like, to charity."

"To charity?"

"Yeah, right. That way you'd be doing a lot of good."

"Listen, it doesn't matter what I do with it, I'm still taking it. Why should I take it?"

"Well, Frank, it would just make everybody feel better."

"You're telling me I should give it to charity. What the hell are you taking it for?"

The plainclothesman reddened. He started to talk passionately about his poverty-stricken childhood. "Nobody gave me nothing. Now *I'm* taking. I'll take anything—two bucks, ten, a hundred, a grand! Whatever it is, I'll take it."

"Look," Serpico said, "what's your education?"

"I got a high-school equivalency."

"And how much are you making a year?"

"The same as you, I guess. About twelve grand. So?"

"With that kind of education," Serpico said angrily, "where the hell else are you going to make money like that, with all the benefits and the pension and stuff? And still you got to take the money? You're a cop, for Christ's sake! You know what that means? You ever stop and think what you're doing? Do you ever lie in bed at night and think about it?"

The plainclothesman looked away. "No," he said finally, "I can't think about it. If I thought about it, I'd take my gun and blow my brains out."

There was a bizarre aftermath to the great meeting about Zumatto. Stanard came to Serpico a few days later and spoke to him about James Paretti, the new man who had taken Zumatto's place in the 7th Division. "Listen, Frank, you know this guy Paretti is from downtown, and he could be a plant. We want you to check him out, see what you can find out."

"You want *me* to check him out? Why me?"

"Well," Stanard said with a certain logic, "you aren't taking any money, and you can talk to him. You don't have anything to worry about, right?"

The other men in the division had been avoiding Paretti, and Serpico had been wondering how he could get close to him to see if he was in fact an undercover man sent by First Deputy Commissioner Walsh. The assignment from Stanard was a perfect opportunity to do so without causing any comment. Serpico simply went to one of the division lieutenants and said he wanted to work with Paretti for the day.

Over lunch Paretti started expounding on the swath he was going to cut through the South Bronx. "The first guy I'm going to lock up," he said, "is Moe Schlitten." Moishe and Samuel Schlitten were the Bronx policy-numbers kings, allied with at least two of the five Mafia families in New York. For cops like Serpico or Paretti, arresting either of the Schlitten brothers was the same as walking in and arresting the president of U.S. Steel or General Motors. The level at which they were paying off was something Serpico could only imagine.

"I'm going to grab Schlitten," Paretti said. "Unless he's on the pad. I guess he is, huh?"

"I wouldn't know," Serpico replied.

"Listen, they send you around to check me out?"

"What makes you say that?"

"Well, what the fuck, nobody's been throwing their arms around me. I figure there's got to be a pad. Why should this place be any different? What's the nut a month?"

"I don't know," Serpico said. "I don't take any money."

"Yeah? Well, what do you hear, say?"

If Paretti was an undercover man, Serpico thought, he was pretty good. "Around eight hundred, that's what I hear."

"Eight hundred, huh? When do you start getting it?"

"The way they talk, it's a month and a half."

"A month and a half!" Paretti said. "The hell with that. I want mine now."

The next day Stanard asked Serpico what he had learned, and Serpico told him that Paretti had questioned him about the pad and wanted it right away.

"Fuck him," Stanard said. "Why should he be any different?"

chapter 12

Throughout the summer of 1967 the psychological crunch on Serpico became almost unbearable. Besides his nerve-racking position in the 7th Division, there was a personal stress on him that in its way was infinitely worse, and had nothing to do with corruption. Against every value of honesty and forthrightness that his upbringing had instilled in him, he found himself living a lie in Greenwich Village. None of his friends and neighbors on Perry Street knew he was a cop.

The metamorphosis in Frank Serpico, the disintegration of his pride in being a police officer, was very nearly complete. Just two years before, for example, the biggest political issue in the city was whether there should be a civilian review board overseeing some activities of the Police Department, and Serpico, like practically every other member of the force, had been against the idea, convinced that police officers could take care of their own affairs without outside interference. Now he was even concealing the fact that he was a cop.

At first it had not been a problem. Life in the neighborhood was casual, people came and went, and nobody bothered to find out what anybody else did. But as Serpico made friends, saw them in coffeehouses, was invited to their homes, he was appalled to discover a nearly universal contempt and fear expressed about the police—not only the usual allegations about police brutality, but about police indifference to calls about burglaries just because they were in the Village, about the sly remarks cops made to girls who reported that they had been molested, about their lack of civility, or professionalism, as Serpico would characterize it, especially to someone whose hair was long or who had a beard.

Serpico deeply believed in his concept of the cop as the community's friend and protector, and at first he put down the complaints as radical talk. But he liked these people; as far as he could tell, they were by and large gentle and exceedingly law-abiding, except that many of them smoked marijuana, and although Serpico did not—he excused himself by saying, "I trip on other things"—he had come to realize that marijuana was not exactly the road to perdition. As time went on, he tended more and more to side with them. His dress and appearance made him indistinguishable from most Village residents, and when he walked down the streets, some of them would smile or nod at him and say, "Peace, brother." My God, he thought, if they knew I was a cop, they'd hate me. Sometimes, too, he would walk by a police officer, and could sense the reflexive hostility. To the cop he was another hippie, and Serpico found it easy to believe what his friends had been telling him, that the cops were out on the street not to protect them, but to get them.

So finally, when he was asked what it was that he did, he said he was a remittance man, that he had never been able to get along with his family and they were willing to pay anything to keep him away. Everyone he told this to accepted the story, and good-naturedly toasted his luck at having such wealthy and wise parents. Occasionally a girl who stayed over in his apartment would spot his gun and would ask why he had it, and he would mutter vaguely about needing it for protection, or that he was "into something" and the fewer questions the better. For a while, as a result, the rumor spread that he was a member of the Mafia.

Whether remittance man or Mafioso, all of this only served to give Serpico a romantic, mysterious aura, and in the beginning he rather enjoyed it. But then the guilt of what he had done dug into him, and as he became more entwined in the lives of his Village neighbors, he started to worry about what would happen if one of them were arrested—for smoking pot, say—and it was discovered that he was a cop. They would conclude, of course, and he could not blame them, that he had been sent to spy on them, to try to pin something on them.

During this period only one person knew about the pressures grinding into him because of his triple life—being a cop, unwilling to overlook the crookedness of other cops he worked with, and denying that he was a cop at all in the Village. She was a nurse, and the first girl Serpico had seriously considered marrying. Indeed he was teetering on the brink of it when his nightmare existence destroyed their relationship. He had met her on one of his vacation trips to Puerto Rico, and then saw her with increasing frequency in New York, introducing her to his parents and bringing her to the Serpico family dinners his mother presided over every Sunday. She did not live in Greenwich Village, so that was no problem, and she was aware from the start that he was a police officer. She was essentially indifferent to the fact that he was a cop; it did not mean much to her one way or another. Serpico was what mattered, and what made him happy, satisfied him, was all she really cared about.

Then when all the tensions of his life as a cop became part of every minute they were together, something that she was forced to share and yet really did not feel, she could not endure it anymore, and fled to Europe to make herself break off with him.

She has since married and moved out of the city, and looks back on that time with sorrow and compassion and pain. She simply was not, she recalled, prepared to cope with the situation. "He was so outgoing and full of fun, not that he didn't have his moments—like if every once in a while I smoked a little grass, he'd get very righteous about it—but when the corruption thing began, Frank changed completely. He became so moody and depressed, and it affected everything we did."

After that came the quarrels between them. "A lot of it was my fault," she said. "I remember how he'd say he couldn't do his job anymore, and how he hated it. I got sick of listening to how corrupt everyone was, and how nobody wanted to do anything about it, and I would yell, 'If you hate it so much, why don't you get out?' and Frank would say that I didn't understand what he was going through, and that he had to talk to someone, and if he couldn't talk to me, who could he talk to? Well, it went on like that, and it was a bad scene. I don't know, but it seems like there were months when I didn't do anything except cry."

She recalled that in his anger about the corruption he had encountered Serpico rode an emotional roller coaster, one moment declaring that he would not rest until he had wiped it out, at another moment sliding into deep melancholy, agonizing over how he could ever possibly accomplish anything by himself, alone. There were other moments, too, she remembered, when he looked "scared, like a little boy."

"The worst thing," she said, "was watching him go to work, dreading it. But he couldn't let go of it either. I guess what he wanted more than anything else was just to be a good cop. You wouldn't think that'd be so hard."

In the 7th Division any lingering hopes Serpico had that Paretti might be an undercover cop disappeared later in the summer during a raid on a numbers bank in a Puerto Rican section of the South Bronx. The raiding party consisted of Serpico, Stanard, Paretti, and a fourth plainclothesman. Usually Stanard kept tabs on the locations of the various banks on the pad, but sometimes they moved, and he was not aware of this one. When they broke in they were confronted by a swarthy, squat man with no neck, just a round head perched on top of his paunchy little body. It was the banker himself, Manuel Ortega, and he was furious. "I'm on! I pay for the month!" he screamed at Stanard. "What the hell you do?"

Serpico had never seen Stanard so nonplussed. "Well," he said finally to Ortega, "we didn't know it was your place." Then Paretti saved the day for him. There was a wall phone, and next to it a sheet of paper with a list of names. Paretti ripped the sheet off the wall. "This must be his collectors. Want to check it?"

"Yeah," Stanard said, and after he had studied the list for a moment his whole demeanor changed. "Hey, Manny, you're paying for twenty people, and there's twenty-six here. That's not very nice, Manny. You been holding out on us?"

Ortega began screaming again that he had been paying and paying, and what good was it? Stanard replied, "Anyway, we'll let it go for now, but we've been watching this place and we have to have a collar, so give us one of your people."

This brought another hysterical outburst from Ortega. "No, no, no," he yelled. "I pay all the time, and I get nothing. *I* go. Lock *me* up!"

"Hey, Manny," Stanard said, "don't get so excited. We don't want to lock you up."

"Lock me up, I say. You want trouble, I give you trouble. I don't care for nothing no more."

Paretti interrupted them. "Look," he said, "you guys go outside, and let me talk to Manny here."

Ten minutes later Paretti joined them in the car, smiling and waving a fistful of bills. He had calmed Ortega, he said, and laid down the law, business was business, and if he wanted extra collectors he had to pay for them, or he wouldn't have any business left.

When they arrived at the 48th Precinct, which housed the division offices, Paretti began dividing the money.

"I don't want any," Serpico said.

"Why not?" Paretti asked.

"Hey," Serpico said, "do I have to go through all that again?"

"OK," Paretti said, "so Frank gets ten bucks apiece for expenses," and as they got out of the car, he stuffed thirty dollars in Serpico's shirt pocket.

In the morning Serpico returned the money, putting a ten-dollar bill in each of their mailboxes, and then, having given up on Paretti, he began to withdraw more and more into himself, as he had done in Brooklyn after he was handed the three hundred dollars.

He spent time on the street by himself, without telling the other plainclothesmen what he was doing, and one day from a rooftop in a predominantly black neighborhood he spotted some booming numbers action in an alley. There was a lookout stationed on the sidewalk in front of the alley, and Serpico, dressed in dungarees and an old shirt, decided to try to slip by him. He picked up a bag of garbage and an empty box, and limped along the street, hoping to pass himself off as a delivery boy. The strategy worked. The lookout barely glanced at him, and Serpico was in the alley. About halfway down it, where one of the alley walls recessed, he could see the collector in a chair, a phone book on his lap serving as a makeshift desk, busily taking bets from a line of players. Serpico dropped the bag and the box, and started for the collector. As he did, he heard the lookout behind him shout a warning, "Bill, rise up! It's the Man!"

Serpico began to whoop and holler crazily, trying to drown the warning out, but the collector heard it and leaped to his feet, scattering money and policy slips on the ground, and ran down the alley. Serpico raced after him, past the startled bettors, through a backyard and into the next street. Finally he yelled, "Stop, or I'll shoot!" At that the collector turned and came slowly toward Serpico.

"Man," he said, "you didn't mean that now, did you?"

"Are you kidding?" Serpico said in mock alarm. "It isn't even loaded."

"Ah, that's my man, I like your style. But I got to tell you, you come out of nowhere on me, like a regular jack-in-the-box."

Serpico walked the collector back to the alley. Betting slips were fluttering all over it, although the neighborhood kids had snatched up most of the money, and as he was leaving with his prisoner and the evidence, the lookout arrived with the next level of the numbers-racket hierarchy in tow, a controller.

"What's going on?" the controller said. "We're on, man. We're cousins."

Serpico had not felt so good in months. He did not care in the slightest whether the controller was telling him the truth, and pushed past him.

"I'm telling you, we're good," the controller insisted. "Check it out. Check it out. Can't you make a phone call?"

"No, I can't," Serpico said.

"Well, listen to me then. That man you got there, that's the best worker I got. How 'bout I give you somebody else?"

"Hey," Serpico said, "you think I'm giving this guy up after chasing him all over the South Bronx? Goodbye."

After the prisoner had been booked, there was a tense confrontation with Stanard. "Cheech, they were on!" he said. "What the fuck are you doing? You want collars, we'll give them to you, any kind you want, but lay off the cousins."

"Listen, I didn't know if they were on or off, and I don't give a damn. I saw a couple of white guys in a car up the street there. They could have been inspectors from downtown, how do I know? Get this fucking-well straight, Stanard. I'm not sticking my neck out for you. You do your thing. I'll do mine."

The estrangement between Serpico and the other plainclothesmen in the division steadily worsened, and the wisecracks about how he was probably a spy planted in their midst began taking an ugly turn.

Earlier in the year, the day before his partner Zumatto was transferred, Serpico had tagged along when he went to see the chief of the rackets bureau in the office of the Bronx District Attorney about a pending vice case. When Zumatto left the office, Serpico stayed behind and spent the better part of an hour discussing police corruption with the rackets chief, Seymour Rotker. The conversation was in general terms. Serpico spoke of the amount of bribery and graft in the department, and how nobody seemed to be doing anything about it. Rotker agreed that there were undoubtedly many crooked cops, but that it was difficult to attack the problem because it required the cooperation of individual policemen, and they would not talk.

Now, as the pressures mounted on Serpico, he spoke to Rotker a second time, on the telephone, and posed the corruption issue again. Suppose a cop who was knowledgeable about it came forward, Serpico said, how far was Rotker willing to go? What Serpico was seeking was some kind of all-out commitment, but Rotker replied that he would have to review the material before he could make any promises, and that simply wasn't good enough for Serpico.

It was then, with his mood blackening more every day, that his girl, the nurse, reached the point when she could no longer stand it, and after a last emotional exchange of recriminations, she flew to Europe to forget him. "I just saw it going on and on, and nothing ever coming of it," she later said. Serpico raged inwardly at her, at himself, at everyone who had rebuffed him. It was, he remembered, the most despondent time in his life.

Finally on Sunday, October 1, in a distraught, impulsive moment, Serpico picked up the phone and called Captain Behan. He told Behan that he had put up with all the corruption he could, and if it meant going back into uniform, that was fine with him. Not only was the pad in the 7th Division bigger than ever, but other plainclothesmen were constantly attempting to involve him in their "schemes," and despite Behan's repeated assurances, he had not heard a peep, directly or indirectly, from First Deputy Commissioner Walsh.

Behan sounded genuinely surprised by this. "Frank," he insisted, "I had no idea."

Well, it was true, Serpico said, and with the division plainclothesmen routinely taking him into places that were paying off, how could he prove that he was not participating in the pad himself if by some fluke it was discovered?

"Captain," Serpico said, "I think it's only fair to tell you that I've been to"—and he uttered two words that would change everything—"outside agencies. And I'm going to more of them if I have to."

The alarm in Behan's voice was immediate. "Outside agencies!" he exclaimed. "What outside agencies?"

"I really don't think I should discuss it with you."

"Frank," Behan said, "you can get into trouble for that. It's against the rules. We can clean our own laundry."

"Yeah, that's what I used to think," Serpico said

scornfully, "but not anymore. I'm not holding anything back, and I don't care who gets hurt, including myself."

Behan's effort to regain his composure was evident over the phone. He urged Serpico not to do anything rash. In the meantime he would check into what was going on and get back to him.

Serpico had hoped that his call might stir Behan into contacting Walsh again. But Behan, perhaps understandably, had no stomach for bracing the feared First Deputy and asking him why he had allowed six months to elapse without having been in touch with Serpico, or if he had done anything at all about the conditions Serpico had reported. Instead Behan went back to his neighbor Philip Sheridan, the 7th Division's administrative officer, and told him that Serpico had found "extensive corruption" among the plainclothesmen, that there was an organized pad, and that Serpico had already gone, or was threatening to go, to outside agencies.

Serpico was horrified when Behan revealed what he had done. Deputy Inspector Sheridan had not only described the division as "clean as a hound's tooth," but had been its administrator for more than a year, and before that had served as its public-morals specialist in direct charge of the plainclothesmen. It was almost inconceivable to Serpico that Sheridan could have been unaware of what was happening right under his nose, and he told Behan so.

"Frank," Behan said, "I had to do something," adding that he had complete faith in Sheridan's personal integrity.

"Well," Serpico said, "I notice you're not saying anything about how smart he is."

The Police Department's military structure was such that during the time Serpico had been in the division, he had probably not spoken more than a dozen words to Sheridan, a humorless, balding man who looked as if he might be at home in a bank. But the very next day there was Sheridan standing in the middle of the office, waiting for him; and in full view of the other men, he came over to Serpico and whispered that the division commander, Deputy Chief Inspector Killorin, wanted to see him when he was finished for the night. Serpico nodded. He felt like asking Sheridan why he didn't use a bullhorn and make a general announcement for everybody's benefit?

The meeting between Serpico and Killorin was not auspicious. Deputy Chief Inspector Stephen Killorin had been in command of the 7th Division for approximately four months. He was a tall, lean man with a face that appeared to have been hacked out of granite. He eyed Serpico icily, and began by inquiring if he thought he was the only honest cop on the force. "What do you think I've been doing all these years?" Killorin said before Serpico could answer. "Why didn't you come to me with this information?"

Serpico stared right back at him. "With all due respect, Chief," he said, "I don't know you from the next man. I've never talked to you before."

Killorin said that his record was clear, that he had never hesitated to break police officers who were "derelict in their duty," and then he demanded to know what outside agencies Serpico had been to.

Serpico was still defiant. "I don't see how that's pertinent," he said—thinking to himself, that's what they're really worried about, that's what has them all shook up.

Well, Killorin continued, what did he know about a pad?

"I don't know any more than anyone else around here," Serpico snapped. "Christ, it's been going on long enough."

Then Killorin asked if he was going to cooperate or not, and Serpico bluntly replied that he would have to think it over. Maybe he was just an ordinary cop, he said in a voice laced with sarcasm, but he did not see how the division could investigate itself.

Killorin tried another tack. He pointed out that Inspector Jules Sachson, who commanded plainclothesmen in the Bronx on the borough level, had recently been with the Police Commissioner's Confidential Investigating Unit. Would Serpico talk to him? What did he think about Sachson?

Serpico remained noncommittal. "I don't have anything personally against him," he said. "I've heard his name mentioned. That's all."

"OK, you can go now," Killorin said, and as Serpico turned to leave, Killorin's manner seemed to soften somewhat. "I want you to know," he said almost apologetically, "there are men in the department who feel the same way you do."

During the week of October 1, Serpico had two more sessions with Killorin, both with Sheridan present. He provided the addresses of four numbers operations on the pad, and also named Victor Gutierrez, the informant whom he had seen being given ammunition by one of the division plainclothesmen, as a key link between the cops and those who were paying off. But he remained hostile and suspicious after having waited vainly for months for some word from Walsh, and he was barely able to hide his contempt for Sheridan, convinced that the pad in the division could not possibly have flourished without the knowledge or at least the tacit consent of its superior officers. And even if they were not directly implicated, he thought, they were at least accountable for the supervision of their men. How could they possibly investigate themselves?

On Sunday, October 8—according to the official inquiry that eventually tried to put together all the pieces of the Serpico affair—Killorin met with Assistant Chief Inspector Thomas C. Renaghan during a New York Giants football game at Yankee Stadium. At the time Renaghan was the Bronx Borough Commander, in charge of some thirty-five hundred uniformed patrolmen and plainclothesmen. Killorin advised Renaghan of what was brewing in the 7th Division, that it could easily develop into a first-class scandal, that Serpico appeared reluctant to cooperate with him, and he suggested that Inspector Sachson might have better luck. Renaghan agreed to let Sachson work with Killorin, but told him that any allegations that turned up should be given to Police Headquarters to handle. Renaghan's apparent disinterest in concerning himself with the problem became more understandable three years later when he was convicted of a felony, criminal contempt, for refusing to answer questions in a bribery case that involved one of the biggest bookmakers in the city. Renaghan's conviction was subsequently reversed, although this is being appealed by the Manhattan District Attorney's office. As another assistant chief inspector in the department observed, "Renaghan wouldn't have touched a guy like Serpico with a ten-foot pole."

The next day, October 9, Serpico was off duty. Inspector Sachson called him at home, said he had to see him immediately, and arranged to meet him not far from Serpico's apartment, at the Nineteenth Street exit of the West Side Highway. Having been forewarned by Killorin about Serpico's attitude, Sachson, a heavyset, ham-handed man, attempted the friendly approach with all the aplomb of an ex-army colonel used to having his way. He cited his own anticorruption efforts with Police Commissioner's Confidential Investigating Unit, and told Serpico he sympathized with the strain he was under, but that all Serpico had to do was pass on the information—what did he have on whom?—and Sachson would take it from there.

"Yeah, sure," Serpico said, "I've heard all that garbage before. But how far are you going to go? What about the bosses? You going after them, or just some flunky cops?"

Sachson then let Serpico know precisely where he stood.

"Look, kid," he said, "it's in our hands now. If you want, we can just pull people in, and tell them what you're saying, and they'll deny everything, and that'll be it. You're going to have to trust me."

Serpico leaned back in the seat of Sachson's car, drew a breath, and decided that at this stage Sachson was right, he didn't have any choice. So beginning with Stanard and Zumatto, he listed all the plainclothesmen that he knew had been sharing in the pad, and, while Sachson scribbled furiously in a notebook, gave him approximately thirty locations in the South Bronx that were paying for protection.

The same day Sachson, Killorin, and Sheridan trooped down to Police Headquarters to confer with Supervising Assistant Chief Inspector Joseph McGovern, then head of the Intelligence Division, and of all the ranking officers in the department the one considered to be the closest to First Deputy Commissioner Walsh. As soon as McGovern learned why they had come, he notified Walsh, and everyone adjourned to Walsh's office.

Killorin did most of the talking, and the Machiavellian sixty-three-year-old Walsh listened, his beetle-browed face, as usual, impassive. Killorin mentioned Serpico's name several times. He said that Serpico "tells us" of the existence of a pad in the 7th Division, that it included personnel on both the precinct and borough levels as well, and that an investigation was obviously in order.

When Killorin had finished, everyone looked toward Walsh. There was a long, uncomfortable silence. Then John Walsh said, "All right, you've got it."

Perhaps a minute passed before it sank in that Walsh was throwing responsibility for the investigation right back at them. Hesitantly, Killorin asked if this was not a matter for headquarters to take over.

"No," Walsh said in what for him amounted to a speech: "You handle it. You're experienced men, and you have enough captains and lieutenants up there to help you."

The officers from the Bronx sat rooted in their chairs, stunned by Walsh's decision, but none of them dared to protest it. Killorin asked one more question. Could the man in direct command of the 7th Division plainclothesmen, a deputy inspector named Benjamin Hellman, be informed of Serpico's allegations?

Walsh said no, too many people knew about them already. He added that Chief McGovern would supply whatever "technical assistance" was required, and dismissed his visitors.

Not once during the meeting did Walsh indicate that he had ever heard of Frank Serpico.

chapter 13

An officer in the New York Police Department can advance to the rank of captain by passing competitive civil-service examinations. From then on, through the various grades of inspector, he serves at the pleasure of the Police Commissioner. This unpleasant fact, while it is not reflected in the official reports following up Serpico's charges, was very much in the minds of those saddled with the investigation.

Serpico ran into it almost at once at a second West Side Highway interrogation conducted by Inspector Sachson and Deputy Inspector Sheridan, after their meeting with Walsh. Sachson was pressing him about some of the locations in the South Bronx that were on the pad. Serpico could not recall a particular address, just the area. Sachson became impatient, and Serpico said, "What do you want, the whole package signed, sealed, and delivered? I've got to think. I'm giving you the information. *You're* supposed to be doing the investigating."

Sachson exploded, "Sure," he said, his voice quivering with indignation, "you drop this hot potato in our laps, and that's it. You got nothing to lose. We didn't ask for this. Today we're inspectors. Tomorrow we could be captains."

Sachson adopted a warmer tone at another meeting of the three men in a room rented under the name of "Mr. Gold" in the Concourse Plaza Hotel in the Bronx. They could all relax here, Sachson said, and just talk, go over the whole thing again. During the conversation, Serpico spoke of a lieutenant who had told him that if he was worried about a place to stash his money, he was welcome to use the lieutenant's attic. He also mentioned another lieutenant who he was sure knew about the pad but was cagey enough not to have said anything directly to him.

"I'll have them transferred, first thing," Sheridan blurted out. This was the traditional way superior officers treated potential trouble in their units: ship a man out, let him be somebody else's problem. But Sachson, perhaps sensing Serpico's rising anger, quickly interjected, "Transfer them nothing. I want their asses too."

By then Serpico had supplied thirty-six locations that were paying off, and there were more meetings with Sachson and Sheridan, sometimes with Killorin. These gradually took on the aspect of an inquisition.

What *exactly* was this address, that one? For more than six months, Serpico on his own had been trying to keep general tabs on the operation of the pad while waiting for instructions from Walsh, fully expecting to become an integral part of an anticorruption team that would develop and widen the investigation. Now one specific after another was being demanded of him. In some instances Serpico was able to provide only a phonetic version of a particular policy banker's identity, and accusatory questions came hurtling in at him—how *precisely* did you spell so-and-so's name? Whenever he asked any details about the investigation, however, he was told that it was not his concern.

Yet at the same time Sachson wanted Serpico to wear a hidden recorder to tape other plainclothesmen in the division. Serpico refused. He reiterated his conviction that by itself this would do nothing to correct a condition that permitted graft to run wild, and he also pointed out that as a practical consideration the division plainclothesmen had grown increasingly wary of discussing the pad with him, that even though they did it supposedly as a joke, they were constantly patting his body and clothes looking for a recorder, and it would simply be a matter of time before they discovered such a device if he actually had one on him.

The meetings Serpico had with Killorin usually took place after dark in a deserted supermarket parking lot, and on the night of October 27, 1967, after he had met Killorin there, another car drove up. A thin, gray-haired, ruddy-faced man joined them. He was Deputy Chief Inspector James Duck, second-in-command in the Bronx. Serpico was required to tell his story all over again, and he remembered how Duck listened coldly, wordlessly, to what he said, and once more he felt that somehow he had become the defendant in the case.

Finally Supervising Assistant Chief Inspector Joseph McGovern himself, the man directly under John Walsh, called Serpico at home, told him he had wanted to get together with Serpico for a while, to hash things over with him, and arranged to see him privately near the Fulton Fish Market in lower Manhattan. Serpico went to the meeting eagerly. McGovern was the highest-ranking officer he had ever met, and he saw it as a chance to convey his passionate concern about the Police Department. Serpico had committed his life to being a cop. What he was seeking was a fundamental change in the system that made corruption a way of life in the department, and he feared that the investigation in the South Bronx would simply gloss over the situation, possibly make scapegoats out of a few plainclothesmen, and let it go at that. When he voiced all this, however, he found McGovern unresponsive. McGovern suggested that Serpico was overreacting. The system was sound. The problem was some rotten apples, a hazard any large organization had to contend with. And when Serpico again attempted to determine the scope of the investigation, McGovern grew vague and counseled him to be "patient."

There was a very good reason for this reluctance to tell Serpico about the supposedly massive effort the Police Department was launching against the corruption he had reported. The investigation was practically nonexistent. It was supervised by Killorin, whenever he could spare the time from running the 7th Division. All the legwork, surveillances, and observations were being conducted by Inspector Sachson and Deputy Inspector Sheridan, who also had their regular duties to perform, together with one captain, one lieutenant, and occasionally a sergeant. Finally, Walsh's instructions to the contrary, the deputy inspector in command of the division plainclothesmen, Benjamin Hellman, was brought into the investigation on a limited, "need-to-know" basis. The "technical assistance" that McGovern was to provide consisted of a couple of undercover cops who made bets at some of the locations Serpico had pinpointed; a recording device called a Minifon that wound up being used mostly to tape interrogations of Serpico; and the installation of one wiretap that proved fruitless.

The undercover cops succeeded in placing bets at two or three locations, but when the operators of these places were questioned by Sachson and Sheridan, they denied paying off the police. Manuel Ortega, the policy banker Stanard and Paretti had mistakenly raided, was also questioned, and insisted that he had no knowledge of a pad. Not until January 19, 1968, three and a half months after the investigation began, was the first tangible bit of evidence uncovered. Sheridan, while searching still another location supplied by Serpico, found the unlisted telephone number of a 7th Division plainclothesman named Philip Montalvan.

Despite the pathetic nature of the investigation,

word that something was going on soon filtered back to the 7th Division plainclothesmen, undoubtedly from the policy operators who had been visited, and Serpico's position became more dangerous than ever. He not only was completely in the dark about the progress of the investigation, but now more than ever was a target of suspicion from the other men.

At first it was just an attitude, something he sensed, plainclothesmen in the division office whispering among themselves and abruptly falling silent at his approach, eyes that followed him and darted away when he turned around.

Finally one plainclothesman made a clumsy attempt to trap him into an admission, accusing him of passing his share of the pad to an anticorruption unit. "I hear," he said, "you're vouchering the money."

This was relatively easy for Serpico to handle. "How can I be vouchering the money," he demanded in sham exasperation, "when I'm not taking it? Who the fuck starts these rumors, anyway?"

Then Stanard took him aside. "If the shit ever hits the fan," Stanard said, "don't think you're out of it because you're not taking the money. You're part of the outfit. You're responsible just like the rest of us."

"How can I be responsible if I'm not taking the money?" Serpico shot back. "What do I have to worry about?"

"That's not the way it works. They don't stop and ask who's who? When they clean out a division, everybody goes."

"Fuck it! I'm not worried. At least *I* know I'm not

taking the money. I can sleep at night. I'll be able to live with myself."

"Think it over," Stanard said, "you're making a big mistake," and walked away.

The investigation limped along for another two months, the probability of its petering out altogether more likely every day, when it suddenly got exceedingly lucky.

One of the locations on the pad that Serpico had given to Sachson was a bodega, a Puerto Rican grocery store, run by a couple named Juan and Dolores Carreras. Serpico had first heard about it from his old partner Zumatto, just before he was transferred out of the division. Zumatto had said he had been out one night with his girlfriend and thought that he had spotted a numbers operation in the Carreras bodega, and was going to check it out. After he left the division, however, Zumatto called Serpico to warn him not to do anything about this information until he spoke to Stanard. When Serpico told Stanard that he was going to look into the bodega, Stanard said, "Listen, don't make waves, they're cousins. You need information for some collars, we'll give you lots of information, but lay off them."

Victor Gutierrez, the slippery 7th Division informant whose name Serpico had also given to Sachson and Sheridan, was questioned about the Carreras bodega, and while Gutierrez was very cute about discussing police payoffs—shifting his story almost daily, trying to figure out where *his* best interests lay—he finally confirmed that the Carrerases were on the pad.

Then an undercover cop was sent into the bodega to make a bet. The bodega was kept under periodic observation and several division plainclothesmen were seen entering the store for, as the official report put it, "no apparent purchase," and on March 29, 1968, Deputy Inspector Sheridan obtained a search warrant.

What nobody in the investigation knew at the time was that the Carrerases, especially Mrs. Carreras, were in an extremely shaky state. They had been running an independent, medium-sized policy game when disaster had struck the previous December. Within two weeks, they were hit by two winning numbers and lost approximately twenty-seven thousand dollars. This could have been sheer coincidence—there were, after all, winning numbers occasionally—but it also could have been arranged to drive them out of business. The most curious aspect of the two heavily played winning numbers was that they had the same digits, although in different sequence. In any event the Carreras operation, unable to continue on its own, was absorbed by a huge Mafia bank based in New Jersey, and the Carrerases were now hired hands, in effect, forced to settle for a percentage of profits that formerly had been all theirs. Dolores Carreras, a dark-haired woman of thirty-seven with a quick, nervous smile, the mother of three children, was frightened by what had occurred and wanted to get out of the numbers racket entirely, but her husband, Juan, refused, insisting that it would not be long before they were back on their feet again.

As it happened, on March 30, when Sachson and Sheridan personally raided the bodega, Mrs. Carreras was there alone. And at this point, still unaware of what was going on, Serpico was finally brought into the investigation. Deputy Inspector Hellman, commanding the division plainclothesmen, had instructed him to remain in the office that day. Then Hellman received a phone call, and said to Serpico, "OK, let's go." Serpico recognized the address Hellman had given him as that of the Carreras bodega. Serpico drove his own car, a BMW sports coupé with a sliding roof, and he remembered how Hellman stood up through the open top, urging him on as if he were leading a cavalry charge. When they arrived at the bodega, Serpico found Sachson and Sheridan grimly on guard, along with Dolores Carreras, a fixed smile on her face, hands trembling.

Sachson directed Serpico to search the bodega, and he discovered a paper bag filled with policy slips behind a freezer, and more in a storeroom. Then Sachson ordered Serpico to arrest Mrs. Carreras and book her.

As he was fingerprinting her in the station house, she asked in a soft Spanish accent, "Why don't you take the money?"

"How do you know I don't?" he replied.

"Because if you were, you would not be doing this."

Instead of being brought to the plainclothes office for questioning, Mrs. Carreras was whisked into another room. News of her arrest, and that she was now closeted with Sachson and Sheridan, spread quickly among the plainclothesmen, and as Serpico went by the office he could hear the buzz of conversation inside. One of the men came after him and asked, "What's it all about?"

"Beats me," Serpico said in some truth. "All I was told was to execute the warrant."

The plainclothesman studied him for a moment and said, "Well, don't say anything, for Christ's sake."

Serpico was furious at the spot Inspector Sachson had put him in. First he had been kept completely out of the investigation, and suddenly, in the middle of it, he was being openly identified with it. But Serpico was wrong about one thing. It was not the middle of the investigation. For all practical purposes, it had ended with the raid on the Carreras bodega, and those in charge wanted to make certain that, whatever else happened, Frank Serpico was directly tied into it, to prevent him from ever claiming that nothing had been done about his allegations of corruption in the Police Department. Later, when Sachson was asked why he had picked Serpico to book Mrs. Carreras, he blandly replied, "It was a good arrest, and somebody had to do it. I felt I was doing Frank a favor."

Dolores Carreras fell apart almost immediately, admitting that her husband had regularly paid off the police and implicating plainclothesman after plainclothesman, among them Stanard, Zumatto, and Paretti. The informant Victor Gutierrez, who had been involved in the Carreras operation, now backed up her statements. The most difficult one to crack was Juan Carreras, and it became clear that if he had been present at the original arrest, the investigation, such as it was, would have ground to an ignominious halt. A tough, dapper, and experienced policy operator, he denied the existence of a pad but finally, faced with the admissions his wife and Gutierrez had made, agreed to talk. And in the middle of May, with the blessing of First Deputy Commissioner Walsh, the matter for the first time was grandly turned over to Bronx District Attorney Burton Roberts. Corruption had been uncovered, the police had investigated, and here was all the evidence, neatly bundled.

Members of the District Attorney's staff were appalled at the slipshod manner in which the police investigation had been conducted. Nobody could know whether it might not have ranged farther and been more effectively directed if the District Attorney had been brought in at the beginning. But it was apparent at once that even trying to obtain indictments against cops with otherwise unblemished records, much less convicting them, would be extraordinarily difficult if the cases were based solely on the word of witnesses who had backgrounds as dubious as the Carrerases and Gutierrez. So everything still hinged on Frank Serpico; if a case was to be made, he would have to testify before a grand jury.

Burton Roberts was a superior district attorney. A stocky, opinionated redhead, then forty-six years old, he had served his apprenticeship on the staff of Manhattan District Attorney Frank Hogan before going to the Bronx in 1966. His theatrical air was amusing to some, boorish to others. He spoke in sort of a perpetual growl, out of the side of his mouth, and his elliptical verbal flourishes—not to mention the cigar from which he occasionally flicked ashes to punctuate a point—was reminiscent of a W. C. Fields monologue.

In an effort to maintain some degree of secrecy,

Roberts first met Serpico in a Bronx motel. Members of his staff were present, as well as McGovern, Killorin, and Sachson. By now Serpico had a fairly good idea of the pitiful extent of the investigation, and he arrived angry and resentful.

At the meeting Roberts did not know that Serpico had spent the better part of a year vainly waiting to hear from First Deputy Commissioner Walsh, nor did Serpico bring it up; he assumed that everyone present knew of it. But Roberts did recognize how disturbed Serpico was, and he tried a paternal tone with him. He complimented Serpico for having come forward—that was "extremely unique," as he put it. Then he asked Serpico to go through the whole 7th Division story. "Make believe," he said, "that I don't know anything about it."

When Serpico had finished, Roberts said, "Well, there's going to be a grand jury on these shitheels, and you will appear as a witness against them."

Serpico thought of the real culprits—the superior officers, the "bosses"—who allowed corruption to thrive in the department and who would escape untouched. He told Roberts that he, and everyone else in the room, could go to hell. The whole thing was a farce. He had come forward under the illusion that his action would result in a broad investigation to clean up the Police Department. "So what happens?" he said. "All they get is some flunky cops, and I end up the schmuck. The bosses must really be having a laugh over this one."

If Serpico was worried about being singled out, Roberts said, he had nothing to fear. Many if not all of the plainclothesmen in the division would be called before the grand jury, and nobody would know what Serpico's actual role was.

"Mr. Roberts," Serpico said, "you don't seem to understand. I'm not afraid to testify, and I don't care who the fuck knows about it—if it meant anything."

Serpico shifted his gaze to Killorin, McGovern, and Sachson, and as the tension mounted, Roberts decided to adjourn the meeting. Afterward one of his assistants asked him what they would do now that Serpico was not going to cooperate. "Don't be so sure," Roberts said. "He didn't say he would, but he really didn't say he wouldn't either."

The news that a grand jury had been convened to look into police corruption was all over the 7th Division. One plainclothesman whom Serpico did not know very well, a fidgety fellow under the best of circumstances, came up to him and glumly asked his advice. "Gee, Frank, with this grand jury and everything, what do you think I ought to do if they call me?"

"You have to make your own decision," Serpico said. "I can't tell you what to do."

"Well, the guys say we got to stick together, and it'll blow over. I guess I better keep my mouth shut and go along with the rest of the guys. What about you?"

"I don't know," Serpico said, "but I'm not sticking my neck out for anybody."

Robert Stanard also questioned him about what he

might say to the grand jury, and Serpico said, "I'll tell the truth. I didn't take any money."

"What else?"

"How do I know what the hell they're going to ask? What I don't know, I don't know."

"Just remember," Stanard said, "this could be serious. A lot of people can get hurt, including you."

"What's that supposed to mean?"

"Think about it," Stanard said.

Serpico was in his car in front of the 48th Precinct when still another plainclothesman slid into the front seat beside him. "What do you hear about the fucking grand jury?"

"How should I know?" Serpico said.

The plainclothesman grunted. "I got a lot to lose. Maybe you don't, but I do. If it ever came to where I could get indicted, and I thought somebody was going to talk about me, I mean before I'd shame my wife and family, it'd be worth a couple of grand to me to have that somebody taken care of."

"Is that right?"

"Yeah."

"Well," Serpico said, "that's your problem."

So it had come to this, he thought. His first reaction was the inevitability of it—that his life might be in danger, not from crooks but from crooked cops. And then he asked himself why was he even making that distinction? Was there any difference between them?

The next day Serpico went to a Manhattan gun shop. As a cop he had always had his .38-caliber service revolver, but it was too bulky to carry comfortably

in civilian dress, and like most plainclothesmen and
detectives he usually worked with his small, snub-
nosed .38 off-duty pistol. Because of its limited accu-
racy and five-round capacity, however, it would not be
much help in the event of real trouble, and in the gun
shop Serpico finally settled on a fourteen-shot, 9-mm.
Browning automatic crafted in Belgium. It was reliable,
accurate, had the firepower he was looking for, and it
balanced beautifully in his hand. Despite its heft, its
flat design allowed him to wear it with relative ease in
a holster clipped to his belt. If the 9-mm. had a potential drawback, it was its knockdown power, but
Serpico solved this by using super-velocity ammunition with lead-tipped bullets instead of ones that were
completely steel-jacketed. The extra powder charge
started the bullet expanding even before it hit its target, and when it did, it would decimate anyone coming after him. Neither the gun nor the ammunition met
with police regulations. It was the least of Serpico's
concerns.

The grand jury was already in session when Burton
Roberts had a second meeting with Serpico in the
apartment of one of his assistant district attorneys,
David Greenfield, late in June.

Roberts was curt. "You've been doing a lot of talking," he said. "Now let's see if you mean it. The time
has come to put your money where your mouth is."

Serpico retorted that he was being asked to become
a "marked man in the department" for nothing. Then
he launched another broadside against the police
bosses, particularly John Walsh.

"Wait a minute," Roberts said, "aren't you laying it on a little about Walsh?" And it was then that Serpico told him how nearly a year and a half ago he had reported the corruption in the 7th Division to Walsh through Captain Behan, and that he had never heard a word from the First Deputy Commissioner.

For a rare moment Roberts was speechless. But then he quickly seized upon this revelation to develop a wily new line of attack. By testifying, Serpico could show them all up. Serpico, he said, kept talking about how only a few lousy cops were involved in the case, but that wasn't so. Much more was at stake. "You," he said, jabbing his cigar at Serpico, "*you* can turn the whole goddamn thing around."

This was only the beginning, Roberts exclaimed, warming to his own argument. He proceeded to paint a sweeping portrait of justice triumphant. Under his direction, the grand jury would probe relentlessly into police corruption wherever it lurked. It would be bigger than the Harry Gross bookmaking scandal! After the Bronx, he declared, would "come Manhattan, Brooklyn, Queens." All Serpico had to do was tell the grand jury what he knew, and it would finish the job. Serpico wouldn't even be needed as a witness in any trials.

Finally Serpico said, "OK."

The next day, June 26, he began a series of appearances before the grand jury, and almost at once he felt betrayed. A grand jury, whether state or federal, is itself an investigative body, and it can go wherever a prosecutor wants to lead it. But the questions which Roberts and Greenfield asked Serpico about the pad concerned only plainclothesmen. None dealt with superior officers.

The grand jury continued to meet throughout the summer and fall of 1968, culminating in November, when Robert Stanard was called before it twice, and not only denied participating in the pad, but said that he had no knowledge of one in the 7th Division.

In the meantime, Serpico retreated more and more into himself. The hostility of the other men toward him increased daily, although there were no overt acts against him or the kind of thinly veiled threats he had received before the grand jury met. It was as if they suspected him, but had nothing specific to back it up.

Then Serpico found out how effective a hidden recorder would have been had he acceded to the endless demands that he carry one. Working alone as usual, not caring who was or wasn't on the pad, he made a felony arrest of a policy operator whose brother was listed by the FBI as a member of a New York Mafia family. Before the case went to court, the defendant's lawyer approached Serpico and suggested that they get together to discuss his testimony. A bribe was obviously in the offing, and Serpico immediately reported the incident, and this time agreed to be outfitted with a "wire." But before he had even had a chance to speak to him again, the lawyer suddenly withdrew from the case. "What can I tell you?" an assistant D.A. said to Serpico. "It was blown."

Early in December, the grand jury nearing the end of its deliberations, Serpico learned that he would be transferred out of the 7th Division. Supervising

Assistant Chief Inspector McGovern himself informed
Serpico of the move. He would be assigned to plainclothes duty, Manhattan North, primarily covering
Harlem. It was a "step up," McGovern said, since Serpico would be operating at a borough rather than divisional level. Serpico was not overwhelmed. After he
had appeared before the grand jury, District Attorney
Roberts had indicated more than once that he probably would be promoted to detective because of his
courage and cooperation. That, of course, had always
been Serpico's dream, and he could not help thinking
about the possibility. But he said nothing about it
when McGovern called and told him where he was
going.

Before he could really brood over his disappointment, however, Serpico made one last arrest in the 7th
Division which so filled him with disgust, and drove
home with such sickening clarity what graft could do
to a cop's self-respect and to law enforcement generally, that all he wanted was to get out of the division,
under any circumstances, as fast as possible.

In his final months in the South Bronx, Serpico had
developed a number of informants, among them a
black youth, a policy runner, whom he could have arrested, but did not in return for information. One day
the kid asked Serpico, "Would you lock up whitey?"

"Sure, why not?" Serpico replied. "I don't care what
color a guy is."

The informant said that the person he had in mind
was "heavy," an Italian who was "mobbed up."

"You give him to me," Serpico said, "and I'll lock
up his ass good."

The informant told Serpico that the Italian was a loan shark and also a top man in a big numbers operation. One of the drops was near a wholesale meat market. Every day the collector for the area placed betting slips behind a loose brick in a wall set back from the street. The operation was so "cocksure," so securely on the pad, that the Italian himself often came by to pick up the betting slips when he was collecting payments on his loans.

Serpico put the drop under observation on December 13. At twelve-thirty P.M., according to his arrest report, he saw "an unknown male"—the collector—"secrete a slip of white paper in the wall of said location." Bearded, and dressed in dungarees and a padded army jacket to ward off the cold, Serpico grabbed a couple of empty cartons and headed toward the drop. To an onlooker he seemed to be just another worker in the meat market. When he got to the loose brick, he removed it and saw the slip filled with betting plays. He quickly initialed the slip, replaced the brick and resumed his surveillance.

Forty minutes later a light-blue Cadillac sedan glided up to where the drop was. Serpico watched as an expensively dressed man of medium build stepped out of the car, adjusted his overcoat, and walked nonchalantly to the brick. Serpico waited until he had taken the slip out, put it in his pocket, and got back into the Cadillac. Then Serpico hurried over and yanked open the door, and said, "Hold it."

The man stiffened in apparent panic, and Serpico cursed to himself as he tried to find his shield in the deep pocket of the army jacket. The man's left hand

started to move. Serpico saw the ever-present diamond ring glittering on his pinky finger, and while
numbers operators rarely carried guns, he thought
for a second that the man might be going for one, and
said, "I'm a police officer," and finally fished the shield
out of his pocket.

The man relaxed immediately. "Jesus," he said,
"why didn't you say so? I thought you were some
fucking junkie, or something?"

"Move over," Serpico said. "I want to get in."

"What's wrong, officer?"

"Hey, come on," Serpico said. He got into the car
and told the man to give him the slip in his right coat
pocket. Serpico examined it, saw his initials and perhaps two hundred plays. Then he took a card case
from the man, identifying him as Rudolph Santobello
of 65-84 Booth Street, Queens, and found three more
slips with several hundred plays. Behind the sun visor
he also discovered eight additional slips representing
controller records.

By now Santobello had a sheaf of hundred-dollar
bills in his hand. When he had counted off six of
them, he thrust the money toward Serpico. "Here," he
said, "let me take care of you."

"I'm not interested," Serpico said. "Keep it up, and
you're going to be in a lot more trouble."

Serpico took him to the precinct involved, the
42nd, and left him inside to be held for booking while
he returned to the car to look for more evidence. Finding none, he went back into the station house, and at
first Santobello was nowhere to be seen. Serpico finally located him in the detective squad room. He

could not believe his eyes. There was Santobello, silk-suited, a diamond tie clasp complementing the ring on his finger, seated at a table drinking coffee and chatting amiably with several detectives and plain-clothesmen.

"Hey, you," Serpico said. "Come here."

"What for?" Santobello asked, an indignant note in his voice.

"What for?" Serpico said in angry amazement. "Just get over here and empty out your pockets."

"I don't have to do that. I'm not going to."

"Look," Serpico said, "you're under arrest. Now empty out your fucking pockets *now*."

"Oh, you're going to get nasty about it."

"I'll show you who's going to get fucking nasty," Serpico said. He stepped forward and slammed Santobello against the wall, and ordered him to keep his hands against it and to spread-eagle his feet, and then began frisking him as he would a suspect in the street.

One of the detectives said to Serpico, "Hey, take it easy. Rudy's a friend of some of the boys."

"Stay the fuck out of this," Serpico snapped. "It's my arrest." He completed the search. "Now you're going to find out what it's like to be a prisoner," he said, and marched Santobello to a detention cage in a corner of the squad room, unlocked the door, and shoved him inside.

The cage was about four feet square. It was already occupied by two filthy, retching narcotics addicts, their eyes and noses running, their bodies covered with sores, and Santobello hastily turned his back to them, his face pressed against the perforated sheet metal, his fingers, poking through the holes, trying to stay as far away from the addicts as he could. He was not saying anything now, nor were any of the detectives and plainclothesmen.

Serpico kept him in the cage for about an hour before he handcuffed him and took him downtown to be photographed. He left Santobello in a detention cell at the photo unit, and went to check his file at the Bureau of Criminal Identification. When Serpico learned what it was, he started to literally tremble with rage. He was still barely able to control himself when he saw Santobello again. "You son of a bitch," he said, "I ought to bounce you off the walls. That's a nice record you got."

Santobello shrank back. "I was a kid then," he said. "I served my time."

"Yeah, sure," Serpico said. Then he telephoned the 42nd Precinct and got the detective who had protested his treatment of Santobello. "This is Serpico," he said. "You know your big pal Rudy, he was in for life, but they let him out after fifteen. You know what they sent him up for?"

"No."

Serpico spat the words out. "For killing a cop! He's a fucking cop-killer! Tell that to all his fucking friends up there."

There was silence on the other end of the line for a moment. "Well, gee," came the injured, oxlike reply, "how were we supposed to know?"

PETER MAAS

chapter 14

In January 1969 the grand jury had completed its hearings and was preparing its report. Grand jury proceedings are theoretically secret, but just before Serpico's transfer to plainclothes duty in Manhattan North came through he bumped into an old friend, a cop he had worked with in Brooklyn, now a plainclothesman, who seemed to know all about them. "Gee, Frank," he said, "the word is you went in there and didn't hold back anything."

"What was I supposed to do," Serpico countered, "stick my neck out for those fucks?"

"Well, everybody is saying that if you got hurt, you would've been taken care of."

"How?"

"Uh, you know, fifty Gs for keeping your mouth

shut. Now nobody's going to get near you. Nobody wants to disturb the system."

Serpico was even less enchanted about being assigned to Manhattan North when he learned that his new commander was Philip Sheridan, who had become a full-fledged inspector—undoubtedly promoted, Serpico jeeringly imagined, because of his ace detective work in uncovering corruption in the 7th Division. When Serpico reported for duty, his reception was exactly as his friend had predicted. It was as if he were a nonperson. There were perhaps a dozen men in the Manhattan North borough plainclothes office, sitting and standing around, drinking coffee and gossiping among themselves. Serpico was elaborately ignored. No one said anything to him or even looked at him.

And then it happened.

As Serpico stood alone, a plainclothesman, Irish, with black curly hair and watery blue eyes, walked up to him. He stopped about three feet from Serpico, and reached into his pocket and took out a knife. He cradled the knife, unopened, in his hand. The others in the room fell silent. Out of the corner of his eye Serpico could see some of them smirking. The plainclothesman with the knife said, "We know how to handle guys like you." He extended his right hand, the one with the knife in it, pressed a button in the handle with his thumb, and five inches of steel blade leaped out, pointing up. "I ought to cut your tongue out," the plainclothesman said.

Serpico tensed. He looked at the plainclothesman, at the knife, at the watching, smirking faces in the

room. It was incredible that this was actually happening inside a police station and he remembered
thinking, They're trying to put me down. He knew he
had to do something, but he didn't know quite what.

Serpico saw the knife move. He did not wait to
find out what the movement was, where it was going.
All his reflexes sprang into action. His left forearm
swung up savagely and chopped against the man's
wrist. The knife clattered to the floor.

In one fluid motion Serpico gripped the man's right
hand across the palm with his own right hand, and
twisted it back and down. The plainclothesman cried
out. Serpico kept twisting his hand back and down,
and the man had to turn and follow it, or his wrist
would have snapped.

The plainclothesman was now completely helpless, doubled over toward the floor, still moaning in
pain and surprise, when Serpico let go of his hand and
shoved him forward with his knee. He landed heavily on the floor, facedown. The knife was a few inches
away from his outstretched hand. Serpico was over
him at once. He whipped out his Browning automatic,
cocked it, and pressed it against the base of the plainclothesman's skull. "Move, you motherfucker," Serpico said, "and I'll blow your brains out."

The man's body went limp, his face was jammed so
tightly against the floor that he could not speak. Serpico kept the gun on him, and looked around the
room. Everyone was frozen in place, and no one was
smirking anymore.

Perhaps thirty seconds passed before one of the

other cops in the room coughed nervously and said, "Jesus Christ, is that a forty-five?"

"No, nine-millimeter," Serpico said.

"Oh, so that's the new Browning, huh? How many rounds does it hold?"

"Fourteen."

"Fourteen? What do you need fourteen rounds for?"

"How many guys you got in this office?"

"Hey, look, we were just joking."

"Yeah, so was I," Serpico said. He stepped back from the plainclothesman on the floor.

The man swiveled his head cautiously and looked up at Serpico and gasped, "I was only kidding. Mother of God, I was only kidding!" Then, as Serpico put the Browning back in its holster, he scrambled to his feet and darted out of the room.

"Um, I guess I better show you where the coffee is," another plainclothesman said.

Inspector Sheridan immediately detached Serpico from Manhattan North for temporary duty in the "pussy posse," the midtown prostitution detail that operated out of the 18th Precinct on West Fifty-fourth Street. That night in a diary he kept from time to time Serpico wrote, "I finally got my big reward for being a good guy. Times Square and the whore patrol!"

Some cops liked the vice detail. The arrest quota was one per man a night. The law required that the girl had to make the initial approach, suggest a sexual act, and mention a fee for it. But many plainclothesmen ignored such niceties; they grabbed the first streetwalker they saw and simply created the dialogue necessary for an arrest in their reports, and were finished work for the night. The whole judicial process provided an easy rationalization for this. Nine times out of ten, an arrested prostitute's pimp had bailed her out by morning, and she was back in business the next evening.

If he was going to be reduced to this, Serpico decided that at least he would play the game exactly as the law said. He would not make any phony arrests no matter how long it took and as a result, the bathroom in his Perry Street apartment began to look like a backstage dressing room, filled with jars of makeup and hair dyes. He constantly reshaped his beard, sometimes letting it grow bushy, other times trimming it back to a goatee. He had a variety of props, especially hats—a great black, slouched affair, a bowler, two or three fedoras, and a beret. He had as many different spectacles, picked up here and there— a pair of horn-rimmed glasses, another pair of wire-rimmed ones, even a pair of pince-nez. He also used his collection of pipes, including a long churchwarden's pipe that proved immensely successful. Occasionally he took along a walking stick; somehow, wary as the whores were, they never associated a walking stick with a cop. To complement his disguises, Serpico sewed foreign labels in a couple of suits, pressed a black eyepatch into periodic service, acquired a collection of used overseas airline tickets, and, since the girls might demand to see some sort of identification, a Swiss passport folder in which he inserted his real passport opened to the page with his photograph. To top things off, he carried an Americana Hotel room key that a friend had neglected to return after staying there once. "I'm confiscating this," Serpico told him, "for police purposes."

Thus equipped, and with his marvelous mimicry, Serpico would sally forth into the night as Max the beer salesman from Munich, Carlos the industrialist from Madrid, Llewellyn the London barrister. One getup that made him look like a refugee from an Amish settlement—a luxuriant beard, the big black hat, and a black overcoat that reached to his ankles— seemed to draw the prostitutes magnetically. But it proved too effective. In short order the word among the Times Square whores was to be on the alert for "The Beard." Serpico learned about it when he was strolling along West Forty-seventh Street toward Sixth Avenue. Ahead of him was a man with an equally full beard desperately trying to connect with one of the prostitutes lining the sidewalk. At his approach, how- ever, the girls scattered like mustangs in a roundup, screaming, "Watch out, it's The Beard! The Beard is coming," and finally, after the same thing happened for several blocks up Sixth Avenue, Serpico saw the man, completely bewildered, hail a cab and ride off. He decided, reluctantly, that he would have to shave off much of his growth.

In observing all the legal niceties of a vice arrest, Serpico could not carry his Browning automatic or even the off-duty .38 that the other plainclothesmen used, since the streetwalkers invariably rubbed their bodies against him, searching for some sign that he was a cop, before they propositioned him. But many of the girls were hard cases who would just as soon assault and rob a customer as have sex with him. So Serpico bought a small .25 pistol which he tucked inside his underwear briefs. If a whore made a move in that direction, it counted as an overt sexual act anyway, and if he was in trouble, he could unzip his fly as though he had other ideas in mind. More than once he had to do just that. In a sleazy hotel room off Times Square, two girls suddenly pulled knives on him and demanded his money. "Well," he replied, casually going for his pistol, "can't we have a little fun first?"

The .25 was so innocuous-looking that one of the girls sneered, "What the hell is that, a cigarette lighter?" and Serpico had to pump a bullet into the bed to convince her otherwise.

While Serpico's credentials, such as his Swiss passport folder, could not stand up under close scrutiny, most streetwalkers, given the slightest excuse, wanted to believe that everything was all right. Some, however, were inordinately suspicious. One night Serpico was on his way down Central Park South. A girl was leaning against a hotel wall. On the other side of her was a delicatessen. As he got closer, the girl went into the store. It was extremely cold, and after he got to where the girl had been standing, he saw that it was over a hot air shaft. He decided to remain there for a while to warm himself. He had almost forgotten about the girl when she came out of the delicatessen and said, "Hey, that's my spot."

She was blond, with gamin features, booted, and, despite the cold, wearing a short coat that showed a lot of leg. Serpico figured she was probably a call girl who hadn't received any calls that night. He often did not know what role he would play until the moment arrived. Suddenly he felt very Continental and became Antoine the diamond merchant from Brussels, and replied in a thick French accent, "Oh, I am so sorree. *Excusez.*"

The girl looked at him with renewed interest. "Where are you from, anyway?"

"I am, how you say, from *la Belgique.*"

"How would you like to buy me a drink?"

"Oh, but of course!"

"Well, we can go right in there," the girl said, pointing toward the hotel bar. Inside, when she had taken off her coat, Serpico looked at the skirt that barely covered her behind and the breasts jutting under her sweater, and began to wish he wasn't on duty.

"What would you like?" she said.

"Some cognac, yes."

The girl snapped her fingers at a waiter and said, "Two Remy Martins," and then said to Serpico, "If you're from Belgium, *vous parlez français.*"

"*Mais oui, je parle français.* How eez it zat you also speaks the French?"

"I'm from Canada," the girl explained. "By the way, where are you staying?"

"Americana," Serpico replied, waving his key, "Americana Hotel."

"And I suppose you have a passport?"

"*Certainement,*" Serpico said, and flashed his Swiss passport folder in the dim light of the bar.

"Could I look at that again?"

Serpico was sure that the girl would see through him now, but then he made an inspired move. He was wearing a lapel button with a picture of Lenin that a friend, recently back from a trip to Russia, had given him as a gag, and he pointed to it, and said, "I do not like zat anyone examine my papers because of my *politique.*"

"Oh, I understand," the girl said.

"Why eez it zat you wish to see my papers?"

"Because I want to make sure you're not a New York City policeman."

"Ooh, la la," Serpico said. "No police, I beg of you."

"Don't worry about it, honey," the girl said. "Now how would you like me to go with you? It'll cost you something."

"How much?"

"A hundred dollars."

"But my dear," Serpico said, "I do not wish to purchase you. I just want to borrow you for a little while."

"Well, you're paying for quality."

"Quality?"

"You know, some *soixante-neuf,* whatever you want."

In the cab the girl told the driver, "The Americana," and Serpico said, "Well, my dear, it is most unfortunate, but zis masquerade must end. I am zee police." He showed her his shield. "And you're under arrest."

It took a moment for this to sink in, and then the girl looked coldly at Serpico. "You're dirty," she said at last, "but you're good."

Serpico worked on the pussy posse for about five weeks when he went on vacation. Before he left, the grand jury returned indictments against eight plain-clothesmen in the 7th Division, among them Stanard, Zumatto, and Paretti, and on his return Serpico would find out why an honest cop who dared cross the crooked ones had to watch his step constantly.

He traveled through Switzerland, Austria, and Italy, visiting relatives of his parents in the town of Caserta outside Naples, and continued on to Spain. In Madrid he happened to pass a shop with a display of knives and went in to look them over. One particularly intrigued him. It was a push-button knife, but instead of the blade snapping open, it thrust directly out. Knives like this were against the law in New York for police and civilians alike, though many cops carried them anyway. With all the thinly veiled threats Serpico had been receiving, he thought it might come in handy, so he told the shopkeeper that he would probably return the next day to buy it. He overslept the next morning, however, and did not have enough time to get back to the shop before catching a plane to London on the last leg of his trip.

After London he flew straight to New York. At the customs booth in Kennedy Airport, Serpico handed over his passport, health certificate, and declaration of purchases. The man checked a list of names on a counter, and Serpico remembered how he suddenly looked a little nervous. He told Serpico to wait, disappeared behind an office door, and returned perhaps a minute later. "Anything wrong?" Serpico asked.

"Oh, no," the man replied.

On the baggage line a second customs man asked Serpico to open his two bags, and started going through them. "Maybe I can save you some time," Serpico said. "I'm on the job."

The man appeared astonished. "Where?"

"NYPD," Serpico said, producing his shield.

"Just a minute," the customs man said, "let me check something."

Serpico watched him walk away and consult with two supervisors. One of the supervisors followed him back and said, "Bring your bags and come with me."

Serpico was led into a small room where two other customs agents stood waiting. He was ordered to strip. He was about to protest, but he suddenly realized what was happening and decided to keep quiet. It was almost an article of faith among other cops that since he lived in Greenwich Village, he must be smoking marijuana or hashish, and that must be what the customs people were looking for. Somehow this had been arranged, but nothing in the customs regulations said a cop could not be stopped, and Serpico figured there was no point in stirring up trouble that might wind up on his record.

So he submitted to a humiliating body search, and his baggage was gone over inch by inch. There was a flurry of excitement when a five-ounce package of TyPhoo tea he had bought in London was discovered. The package was emptied, but all it contained was tea. Finally one of the men turned to the supervisor and shrugged his shoulders, and the supervisor glared at Serpico and stormed from the room.

On his way out of the customs area Serpico spotted the agent who had initially poked into his bag on the inspection line. "What was that all about?" Serpico asked.

"Honest to God, I don't know. I told them you were a cop, and they just said they knew it."

Serpico told him about the push-button knife he had almost purchased in Madrid, and the customs man said, "It's a good thing you didn't. If they were looking to hang you, and I guess they were, that'd be plenty."

Serpico had to admire the subtlety of it all. He could have been thrown off the force if somebody had wanted it badly enough, to say nothing of how it would have discredited him as a witness against the 7th Division cops facing trial.

He returned to the Times Square prostitution detail for about a week and then was transferred back to the Manhattan North borough plainclothes office. After the confrontation with the plainclothesman who had pulled the switchblade, everyone figuratively tiptoed around him, but he remained completely ostracized.

He had nearly reached the breaking point. He felt he had been denigrated by his assignment on the pussy posse, and the customs incident did not improve his mood. But more than anything he was depressed beyond measure by the grand jury indictments. After all the promises of a thorough investigation by the Police Department and all the reassurances that the Bronx District Attorney's office had given him, only eight plainclothesmen, eight lowly cops, had been indicted. There was nothing said or done about superior officers, or about the system. After all he had gone through, eight cops were ticketed to take the rap, and the systematic corruption would go on as if they had never existed; and as far as other policemen were concerned now, he might as well be wearing a leper's bell.

Then Serpico met Inspector Paul Delise.

Philip Sheridan had resigned from the department—probably because his nerves couldn't take me being around, Serpico thought when he heard about it—and Delise was the new commander of the Manhattan North plainclothesmen. He was a short, powerfully built man with a soft voice and hard brown eyes. The first time he saw Serpico in the office, standing apart from the rest of the men, with his beard and shaggy hair, army shirt, dungarees and sandals, an old army gas-mask bag over his shoulder to carry his .38 service revolver since he had his Browning automatic in his belt, Delise thought he must be an informant, and made a mental note to tell his men not to display their informants so openly.

The next day, when Delise saw Serpico again, dressed the same way, still apart from the others, he asked one of the men who he was. "Oh, that's Serpico. He's in 'clothes.' He's been down on the pussy posse."

Delise went over to Serpico and introduced himself, and asked him to come into his office. When they were alone, Delise said, "Are you by any chance related to the Frank Serpico in the 7th Division?"

Serpico stared guardedly at Delise, wondering what this was all about, what new indignity was coming. Finally, defiantly, he replied, "That's me. I'm Frank Serpico."

Inspector Delise got up, walked around his desk, and reached for Serpico. "I want to shake your hand," he said. "You know, you did a wonderful thing for the department. You're a credit to it, and you've got a hell of a lot of guts. I'm lucky to have you in my command, somebody that I can rely on. You're like a breath of fresh air for me."

For a moment Serpico thought this was a crude putdown, but he looked at Delise more closely and realized that he was absolutely in earnest. Serpico felt a surge of exuberance. No superior officer had ever spoken to him like this before. He had never heard of Delise, but he would subsequently learn that he had a reputation for being so straightforward and honest that other policemen sneeringly referred to him behind his back as "Saint Paul."

Delise asked Serpico if there was any one man in the command that he wanted to work with, someone he would feel comfortable having as a partner.

Serpico laughed wryly and said, "It's not a matter of who I want to work with. It's who wants to work with me."

"Well, if we can't find anybody," Delise said, "I'll work with you myself."

He was as good as his word. It was almost unheard of for an officer of Delise's rank and age—fifty—to be traipsing on roofs, creeping down alleys, climbing over backyard fences, but Delise did it with Serpico, and he seemed to enjoy it. Serpico would worry about him sometimes, and Delise would say, "I'm fine. I feel like a cop all over again."

They worked together throughout much of the

spring and summer of 1969. Delise was the first superior he had ever respected enough to call "boss," and Serpico always remembered this period as his best time as a police officer. He let his beard and hair grow even more unkempt, and to reinforce his appearance as a junkie on the street, he often smoked a black Italian cheroot on an empty stomach—which actually gave him a dazed high, as if he were coming off a narcotics fix. The usual procedure he followed was to spot a policy operation, observe its activity from a rooftop with binoculars, and then have Delise back him up when he hit it, swooping down through a skylight, charging up a flight of cellar stairs, blocking off an alley. Another of Serpico's stratagems to avoid attention as he went from neighborhood to neighborhood in Manhattan North was to munch on oranges and grapefruit, like a vagabond on the move, and all Delise had to do for a rooftop rendezvous was to follow the rinds that Serpico left in his wake.

Delise and Serpico discussed the possibility of corruption existing in Manhattan North similar to that in the 7th Division, but their reputations so completely cut them off from the rest of the men that they were unable to pin anything down.

Then one day Serpico reported to Delise that he had zeroed in on what looked to be a major numbers location. It was in a store, the windows painted over so no one could look inside, on the ground floor of a six-story building off Lenox Avenue in Harlem, and it was important enough to have a lookout posted on each end of the block equipped with walkie-talkies, and a third man, also with a walkie-talkie, directly in front of the store. The street was completely inhabited by blacks, and even with his disreputable appearance Serpico doubted if he could get by the lookouts without being noticed. From the roof where he had been observing the operation, he had seen a cellar door. It was in an alley next to the store, but the lookout in front of it often walked by the alley and could easily spot anyone coming along it. Serpico told Delise that the best way to bust the location was to come down from the roof through the building and out the entrance. From there it was only about twenty feet to the door of the store, and they could take it by surprise.

Delise agreed. The two men climbed to the top of a tenement on a parallel street and laboriously made their way across the rooftops until they reached the building where the store was. Serpico peered over the edge of the roof and saw that the situation was still the same, the posted lookouts, people streaming in and out of the store to make their bets.

The plan was for Serpico to proceed down the stairs first, with Delise following in case of any trouble. But when Serpico came out of the entrance of the building, the startled lookout was staring right at him, and recognized him from another numbers arrest he had made not two weeks before. The lookout shouted, "Close up!" and dashed into the store, and by the time Serpico got to the door it was bolted. In his haste Serpico had dropped his police walkie-talkie, and it lay in pieces on the sidewalk. He surveyed it, and the bolted store door and, in a sudden rush of anger and frustration, picked up an ash can and hurled it through the store window and went in after it.

Inside some men were already beginning to burn records, "Police, hold it!" Serpico said. He had his hand in the gas-mask bag where he carried his .38 service revolver, gripping it, but not drawing it. Several startled numbers players huddled back against the wall.

A moment later Delise joined him. The man in charge of the location was what the police termed a "big" collector. He had three assistants writing up the betting slips, and he looked on sullenly while Serpico gathered up the slips for evidence. Very little cash was found, but this was not unusual; in a large operation the money was moved out as fast as it came in.

Almost at once the collector's controller arrived in the store. "Man, what's going here?" he said. "Where you from?"

"The borough," Serpico said.

"The borough!" the controller said, his voice rising in an indignant shriek. "I don't believe it. I just paid the borough." The controller demanded to see Serpico's shield number, and when he was shown it, he muttered again, "But I paid the borough. What kind of shakedown is this?"

"Who'd you pay in the borough?" Delise quietly asked.

The controller started to answer, then stopped and looked curiously at Delise, and said, "Nobody. I didn't pay nobody."

"OK, let's go," Serpico said to the collector.

"Wait a minute," the controller protested. "He's my man. You got to take somebody, take me."

"You want to go, you can go," Serpico replied, "but he's going too."

"I ain't going nowhere, then," the collector said. "You want us, you got to drag us through them people out there."

Delise tried his own walkie-talkie to get help, but it wouldn't work inside the store. "Listen, boss," Serpico said. "I think I can get out through the cellar." So while Delise held the men in the store at bay with his gun, Serpico ducked through the cellar door to the alley, made his way across a courtyard to the street, and called for assistance from a phone booth.

The crowd dispersed as soon as it heard the radio-car sirens, and Serpico and Delise took their prisoners out. But they could extract nothing further from them about police payoffs. Still, with this first direct indication that there was a pad in Manhattan North, Delise and Serpico did some additional probing and concluded that it was far more sophisticated than the one in the 7th Division. The best Serpico could glean from some of his informants was that a "collection agency" of retired cops made the actual pickups, so that none of the plainclothesmen were physically involved. Beyond this, isolated in his own command, Delise could do little, and finally he called Supervising Assistant Chief Inspector McGovern for investigative help. McGovern was away at the time, and Delise got one of his top assistants on the line. Delise explained his problem, and was told that he would have to go through the Manhattan North chain of command for a request like this.

"Maybe I should just broadcast it over the radio," Delise shot back.

"What's the matter," McGovern's assistant said, "don't you trust your bosses?"

Delise did not reply. A few days later, however, he decided to see what would happen if he went through channels. He spoke to the acting borough commander, a deputy chief inspector named Fred Catalano, and repeated his suspicions of corruption among the Manhattan North plainclothesmen. Catalano told Delise that if he needed any investigative help, he could requisition some recruits from the Police Academy.

"Recruits?" Serpico said when he heard this. "You're kidding."

"That's what he said," Delise replied. "We'll just have to go on the way we were."

Serpico kept knocking off numbers locations as fast as he could find them, and as had happened in the South Bronx, one of his black informants asked him if he was only after blacks, or would he also arrest white policy operators? Serpico said, "Try me," and the next time the informant gave him a name and location Serpico knew he was being tested. The collector the informant had supplied was an Italian, Peter Tancredi, and the address was at Second Avenue and 116th Street in East Harlem, where the remnants of a once huge Italian community, which had spawned such famous Mafiosi as Thomas (Three-Finger Brown) Lucchese, Frank Costello, and Joseph Valachi, still held sway and was left virtually untouched by the police.

Serpico immediately put Tancredi under observation, saw him station himself openly on the sidewalk in front of an Italian social club, and watched as people regularly came up to him, whispered a few words, and continued on. Each time this happened, Tancredi would disappear inside the club and show up on the sidewalk a minute or so later. Satisfied that he was taking numbers bets, Serpico was ready to arrest him. The day he picked for the arrest, however, Delise was unavailable, so Serpico brought along a black policewoman to back him up. In a way, he thought, she would be less noticeable on the street if there were any lookouts that Serpico had not spotted. He himself had trimmed back his beard, and was wearing slacks and a sports coat to fit more into the neighborhood pattern. He was so leery of a leak about what he was planning that he did not even tell the policewoman about it until they were half a block from the social club. He instructed her to remain in the doorway of a bank on the corner, and to follow him wherever he went.

He waited until a player came up to Tancredi. The moment Tancredi turned to go into the club, Serpico was after him. He entered the club just as Tancredi was going into a kitchen in the rear. All around Serpico as he walked through the club were elderly Italians playing cards and drinking coffee; none of them looked up, at either Tancredi or him.

At the kitchen doorway Serpico saw Tancredi take a slip of paper out of the oven and start to write on it. Serpico stepped forward, grabbed the slip, took more

of them out of the oven, and said, "You're under arrest." Tancredi looked at him, speechless.

Serpico brought him back into the front room. The
old men playing cards were still paying no attention
to either of them. The policewoman was waiting by
the door, and Serpico told her to lock it. This finally
caused a stir among the cardplayers and coffee
drinkers. The man behind the coffee counter suddenly announced that he wanted to leave. "Stay where
you are," Serpico said. "I want to look around a little
more."

The counterman said, "You can't keep me here."

"Don't pull Philadelphia lawyer on me," Serpico
replied, "or I'll break your chops. There's gambling
going on in here, and what's that guy over there doing
with a beer? You got a license to sell beer?" The man
shut up and started polishing the counter.

Just then there was a tap on the glass street door.
Two men were standing there, one taller than the
other, both expensively dressed and sporting wide-
brimmed hats. Serpico told the policewoman to open
the door. Both men started to enter, the shorter man
first. "Just you," Serpico said.

The shorter man nodded to his companion and
came inside, and all the old men in the club became
very quiet. "What can I do for you?" Serpico said.

"Well, why don't we have a soda or something?"

"I'm not thirsty."

"Hey, relax," the man said. "Don't you know who
I am?"

"No."

"Think about it. I got the same name as you."

Then Serpico knew who the man was—Frank Serpico, alias Farbi, a known gambler in the New York police files and listed by the FBI as a lieutenant in what at the time was the most powerful Mafia clan in the country, the Vito Genovese family. Frank had never encountered a Mafioso of such rank face-to-face, and he wondered how the racketeer had learned about the arrest in the social club. Had the news spread that rapidly? Even more fascinating was the fact that the other Serpico knew who *he* was.

"Can we go back and talk?" the other Serpico said.

"All right." He told the policewoman to keep the door locked, and not to let anyone in or out.

In the kitchen, Serpico the racketeer said, "Look, what's the problem with you, kid? Everybody's OK, and then you come along. What's the matter, huh? I tell you what. I'll take care of you out of my own pocket."

"I'm not interested in what's in your pocket. If I was, I wouldn't be here."

"Listen," the racketeer continued, "you take that guy out of here and you're going to be sorry. He's just trying to make a living. It's bad. It's bad for the whole neighborhood. A thing like this gets everybody upset. I know about you. I mean I know about you and the Bronx and all that, and I don't get it. What the hell, we're the same blood. I know where you grew up. I know the car you drive. I know you live in the Village." He paused, and said, "I even know where your family lives. How are they, your mom and pop?"

There were two ways, Frank Serpico thought, that he could interpret this. His namesake had chosen his words very carefully. Ostensibly he was simply pointing out that they ought to be friends, but there was an implicit threat in everything he had said.

Serpico decided to take it the wrong way. He suddenly drew his Browning, and pointed it at the racketeer's stomach. "I'm through listening to all this bullshit," he said. "I made an arrest and the guy's going."

The Mafia lieutenant flushed. "Hey, come on. Can't we talk to you? What kind of a guy are you? There're other honest cops, but at least they honor the contracts."

"Either I walk out with him now," Serpico said, "or the whole bunch goes. It doesn't make any difference to me."

"OK, have it your way, but I don't think that's the way to do things. It's not nice."

"You got anything else to say?"

"No," the gangster said. "I'll see you around."

It was not long after this that another white numbers operator Serpico had arrested, Vincent Sausto, alias Mickey McGuire, turned to him in court and said, "Hey, you know they're going to do a job on you," and when Serpico asked him who he had in mind, Sausto said, "Your own kind," and Serpico asked, "What do you mean my own kind, the Italians?" and Sausto replied, "No, cops!"

chapter 15

In late 1969, while Serpico was still working under Inspector Delise in Manhattan North, and Bronx D.A. Burton Roberts and his staff were beginning to prepare trial cases against the plainclothesmen the grand jury had indicted, Roberts had an embarrassing problem. Despite his earlier assurances to the contrary, he knew that he would have to fall back on Serpico as his key witness, and he was chary of taking Serpico head on again without some help. So he called upon Assistant Chief Inspector Sydney Cooper to soften Serpico up, hoping that Cooper's hard-driving manner and his tough, anticorruption stance would impress Serpico and help make up for the failure to deliver on the grandiose guarantees to clean house in the Police Department.

Cooper was then borough commander in the Bronx. When he had been assigned to the post in August 1968, he was faced not only with the scandal in the 7th Division, but with a borough on the verge of a racial explosion and last in effective law enforcement in the city in almost every crime category. Since the police phase of the 7th Division investigation was over and Serpico's charges were in the hands of the grand jury, there was not much Cooper could do about them, and he devoted practically all of his attention to working with community leaders on the racial situation and to doing something about crime in the Bronx. By the time he did meet Serpico, racial tensions in the borough had eased dramatically, and the Bronx was outdistancing other areas of the city in combating such crimes as assault, robbery, burglary, and larceny.

Serpico, of course, had heard about Cooper, and Delise once said to him, "If the department had someone like Cooper in charge, the rest of us wouldn't have to be hanging on, hoping for the day men like you would come along and speak up."

Still, the initial encounter between Serpico and Cooper was stormy. Serpico told Cooper that there was nothing personal in what he had to say, that he had the utmost respect for him, but that Cooper was coming into this "after the fact," that he had no conception of what Serpico had been through. The great investigation he had been promised simply had not happened. Why should he go out on the limb even further than he had, Serpico shouted, just so Roberts could convict a handful of plainclothesmen while everything else remained the same? "Who gives a fuck about me?" he exploded. "Except for Inspector Delise, I don't have a friend in the department."

Then Cooper launched into the long, melodramatic scene that Serpico would always remember— Cooper roaring, "Friends! Don't talk to me about friends. Don't give me that bullshit!" and banging his fist on the table, citing the lonely life he had lived as a hated anticorruption commander, and yelling, "I don't have any friends in the department either, and I'm not looking for any," and finally Serpico, unable to keep the smile off his face despite the strain he was under, saying, "OK, Chief, I'll tell you what. *I'll* be your friend."

Possibly things would have been different if Cooper had been on hand when the investigation started, but he had not, and so on December 8, 1969, when Burton Roberts called Serpico at home and told him that he wanted him to be at a meeting that day to discuss his appearance as a witness in some of the trials, Serpico refused, demanding once again to know why police bosses had not been indicted.

Roberts said that there just wasn't enough evidence. He was able to obtain indictments against only the plainclothesmen because of the testimony of the numbers banker Juan Carreras and his wife, Dolores, and the informant Victor Gutierrez. "What you do," Roberts said in one of his florid pronouncements, "is corroborate these witnesses. You add the condiments to them. They are people who admittedly aren't going to win the man- or woman-of-the-year award for honesty, credibility, or good citizenship. But you are different because you are a heroic member of the New York Police Department. So what's the problem? Be here."

Serpico still said he wouldn't come, and then Chief Cooper got on the phone. "Look, Frank, you want me to send out an order? You heard the D.A. Be here."

Besides Roberts and Cooper, Deputy Chief Inspector Killorin, the 7th Division commander, and Inspector Sachson were present at the conference. "Take your coat off," Roberts said to Serpico when he arrived, "and sit with these three officers who are the brass of the Police Department. We are about to engage in the preparation of this matter *today*—not tomorrow, next week, but today. Now I want to make it clear that you deserve a gold shield. So listen to me. Stop being a prima donna. You're a good cop. Don't make yourself a prima donna, or a character. Don't make yourself flaky. Stop the *shtiklech!*"

"I'd like to be a good cop," Serpico retorted, "if I was ever given the opportunity."

"Well, you now have that opportunity."

"There's another thing," Serpico said. "You hold up that gold shield in front of me like a limp carrot."

"Limp? What do you mean?"

"Limp because it's not even a fresh carrot anymore."

Sydney Cooper interjected, "Your name was submitted for it last October. You're on the top of the list, even if you don't care."

"That shield," Serpico said, "is like tinsel the department hands out to the undeserving."

"Well, we're getting very literary here," Roberts said. "What does that mean?"

Serpico said that he had a very good example of what it meant. A plainclothesman, while still under investigation in the South Bronx, had been assigned to the Detective Division. There was an uneasy break in the dialogue until Inspector Sachson said, as if it were a perfectly sound explanation, that the man had a "hook"—an influential contact in the department.

"Sure," Serpico replied, "the bosses always take care of their own."

"Listen to me," Roberts said, "will you listen to me? You're an intelligent guy. You're asking for the complete reformation of an agency of thirty thousand men, an agency which in the past—and I believe I'm older than you—was absolutely replete with corruption. . . . We know the situation. It was a horrible situation. But it's changed. You wouldn't have been able to sit down and discuss something like this with these three fine commanding officers."

"The situation hasn't improved," Serpico said. "It's just gone underground."

"Look," Roberts said, "I'm a great believer in steady progress. I don't believe in going up two, three, four steps at a time."

"Don't give me that," Serpico retorted angrily. "Maybe I'm only a cop, but this was definitely a case for Internal Security. Instead, it was handled on a local level with limited personnel and equipment, and where it ended up is ridiculous. They got a lucky shot in the dark, and now they want to smooth things over, and I'm the key witness."

"Frank," Chief Cooper said, "if we don't convict these guys, everyone will think it was just a big show, that you were just putting on a big show."

"Right," Serpico said, "I'm the fall guy. If I had been given the assignment to investigate, I would have gone out and done it. I was deceived into thinking I was going to be working for Walsh. Now after the whole investigation gets botched, after I give certain information and it doesn't work out, they want me to testify. What's the most we're going to get out of this?"

Roberts declared that he was determined to convict at least "two cops"—Stanard and Zumatto. "It's not important what they get," he said. "It's the fact that two cops were thrown off the force and lost their pension. We're going to show everyone that this shit won't be tolerated in the Bronx."

Remembering the assurances Roberts had given him prior to his grand jury appearance that the investigation would spread to other boroughs, Serpico snapped, "But meanwhile shit like this is happening in Brooklyn and Queens and . . ."

"You keep mentioning Brooklyn and Queens," Roberts interrupted. "Why don't you mention San Francisco and Los Angeles as well?"

"All I'm mentioning, all I care about, is the New York Police Department. You told me that this was going to be bigger than the Gross case."

Even Serpico had to laugh when Roberts replied, "There is no doubt that at times I exaggerate."

"Mr. Roberts," Serpico said, "I'm not saying you didn't do the best you could, and I have the greatest respect for Chief Cooper. But I don't feel any civic responsibility to testify. I feel I've been let down by the department and the whole city administration. They don't give a fuck. They want to lock up a couple of flunky cops and forget about it."

Frank Serpico stared at the faces around him—at the pained countenances of "reasonable" men trying to deal with an unreasonable, petulant child. He knew exactly what they were thinking. But if he had been reasonable in the first place, even the little that had been accomplished would not have taken place. None of them, indeed, would be sitting here now, and there yet lingered in Serpico a hope that something significant would emerge out of all this, that somehow he could prod them, shame them, into really meaningful action.

Finally Roberts requested that Serpico, regardless of what ultimately developed, at least begin working with the assistant district attorney on his staff who was going to prepare and try the cases. Serpico glanced at Cooper. "You've got to do it, Frank," Cooper said, and Serpico, after a long resigned pause, said, "All right."

If Serpico had come away dispirited from the meeting with Roberts, he was ready to throw up after one he had soon afterward with Supervising Assistant Chief Inspector McGovern, who asked to see him privately again at the Fulton Fish Market. By now the various anticorruption units in the Police Department had been completely reshuffled, and Walsh's protégé McGovern was in direct command of them. Although Serpico had seen McGovern a couple of times since the 7th Division investigation had begun, and had talked to him on the phone, this was the first such rendezvous since they had met at the same place more than a year ago.

Serpico drove to the market in his BMW coupé, waited for a while, and was almost ready to leave when McGovern showed up at last, blaming the traffic for his tardiness. A solidly built man with graying hair that had once been blond, and a square, tough cop's face, he had a voice that seemed out of character with his physical appearance; it had an oddly neutral tone. After some chitchat about foreign cars, McGovern blurted, "Frankly, I don't know what I'm doing here, and I think maybe you have the same feeling. But Commissioner Walsh wants to know how you are."

The invocation of Walsh's name seemed to give McGovern renewed purpose, and he hastily added that of course he, too, was concerned about Serpico. He hesitated for a moment. Then, as if he could barely get the words out, he said that he was anxious to create "more awareness" in the department of the problem of corruption, "more enthusiasm" about combating it, and that he hoped Serpico would help.

"I'm concerned about what you feel," McGovern said. "I know you're somewhat of a loner, but everything you've told us so far has a basis in fact. Some things we could do something about, some things we couldn't. I'm looking for intelligence, rumors, some places to start, anything. Frankly, I'm looking for facts."

This sudden, fumbling interest in him after so long

a time meant to Serpico only that the department's high command must be worried about what else he might do, and that McGovern had been saddled with the chore of finding out.

"Look," Serpico said, "I would have put in my papers already if it wasn't for Inspector Delise. I was totally disheartened until I met him. But what's the department doing? There are a lot of good guys out there, but they're just not going to come forward unless they know where they stand. All they're looking for is a grain of encouragement, and they're not getting it."

McGovern's irritation was evident. "I don't think we're communicating," he said. "You're writing a treatise on humanity, and I'm trying to get some facts."

All right, Serpico said, he would stick to facts. What about Captain Foran and the three-hundred-dollar envelope? Why hadn't anything been done about that? McGovern, his lips compressed, chose not to respond. Serpico went on. What about the request for investigative aid which Inspector Delise had made of McGovern's office regarding corruption in Manhattan North, and the advice he received to get some Police Academy recruits? Was that the way to create more enthusiasm about combating the problem?

At the mention of Delise all of McGovern's pent-up resentment seemed to boil over. He had not been in the office when Delise called, but if he had, the answer Delise got would have been the same. "If Delise has a problem," he said, "it's his problem." Delise had been around. If he wanted to get rid of a man, he knew how to "write up a report" to have him transferred.

Serpico started to ask McGovern how this sort of tactic was going to solve anything, but he stopped, sensing that it would lead to endless argument. Instead, he simply said, "It's the hope of the city to back up men like Delise."

Had McGovern not taken this as a personal attack on himself, perhaps Serpico would have let it go at that. After all, it wasn't the first time he had had to listen to a mealymouthed defense of the *status quo*. But he listened dumbfounded, as the man who was supposed to be the Police Department's guardian against corruption smugly declared that, while Serpico might not see it his way, he, Joseph McGovern, had done "a lot"—at least *he* had protected the Police Commissioner "against the onslaughts of outside agencies."

On that note they parted. But McGovern's pointed remark about "outside agencies" especially infuriated Serpico, and he began to reconsider an idea originally proposed to him by David Durk. During the 7th Division investigation he had seen Dirk periodically and had told him of his growing fears that it was going to be a washout. Durk replied that he had the perfect solution. No longer would they try to deal with officials in the city; Durk had a contact on *The New York Times*, and they would go to him and blow everything wide open. Serpico had turned down the idea. He had had enough of Durk-inspired projects—the meetings with Captain Foran, Jay Kriegel, and Commissioner Fraiman. Besides, he reasoned, it was highly unlikely that *The Times* would act simply on the say-so of two cops at their level, even though Durk was a detective. But after the meeting with McGovern, certain now that nothing would ever change with men like him in power, Serpico had another thought. Suppose a superior officer, a full inspector, a Paul Delise, accompanied him to *The Times* and confirmed what he had to say about corruption in the department and the system which allowed it to flourish? That, maybe, *would* make a difference.

When Serpico approached Delise and asked him if he would consider going to *The Times* with him, Delise looked away for a moment and rubbed his cheek, and said, "Frank, I have twenty years, or whatever, in the department, I have a wife and kids, and I just bought a house and there's a mortgage on it, and if I had to leave the department I don't know what other field I could go into. . . ."

Delise's voice trailed off, and Serpico thought, I can't really blame him. He was asking a great deal of Delise, to put his whole career on the line. It was against regulations to do what Serpico had suggested, and if Delise went along with it, the department could throw the book at him if it wanted to. Serpico remembered, too, how other policemen had quietly come up to him, or called him, and complimented him on his stand, but always added that of course they could not be quite so independent and risk so much—they had families, wives and children, to support and worry about. Serpico had largely accepted this, and wondered if he had unconsciously avoided the conditions—marriage and the rest—that they used as an excuse not to get involved. Durk was married and had children, but Durk was different. Serpico had never truly considered him as just another cop; he did not doubt in the slightest Durk's concern about corruption, but he felt that Durk was anxious to make a dramatic name for himself, possibly as a prelude to going into politics, or to rising high in police circles in New York or elsewhere. Once, when Seattle was rocked by a police scandal, Durk had told Serpico that he was in line to become head of Seattle's police force and that he would make Serpico his chief of detectives. Later Serpico said to him, "I'm glad I didn't pack my bags."

Inspector Delise, however, had already done more than his share, more than Serpico had ever expected from any superior officer, and if going to *The Times* was asking too much, Serpico would understand it without question.

The two men looked at each other, and then, as Serpico turned to leave, Delise quietly added, "Well, I made my little speech. Now you do what you have to do. I'll back you up a hundred percent. Anything you want."

Durk's contact on *The New York Times* was a slender, intense, prematurely balding, thirty-seven-year-old reporter named David Burnham. Before coming to *The Times* in 1967 Burnham had served for two years in Washington as assistant director of the President's Commission on Law Enforcement and the Administration of Justice. Burnham worked for the city, or metropolitan, desk as it's called at *The Times*, and his initial assignment was to cover the whole local law-enforcement scene—criminal justice, the court system, jails, judges and district attorneys, the police, and crime itself. He found himself, however, concentrating increasingly on police matters, and he more or less inherited Durk from another *Times* reporter to whom Durk had been feeding tips from time to time.

As Burnham continued to pay particular attention to the police, his indignation about the inroads that graft and corruption were making in the department was based less on moral grounds than on the practical problems of law enforcement, and one of his first major stories about cops—which dealt with cooping, or sleeping on the job—reflected this attitude.

Burnham was already trying to dig out specific evidence of police corruption when on February 12, 1970, Serpico, Delise, Durk, and a fourth cop Durk brought along, who still insists on anonymity although his contribution was minuscule, arrived at *The Times* for an interview conducted by Burnham; Arthur Gelb, the metropolitan editor; an assistant editor; and another reporter.

Serpico's instinct that Inspector Delise's presence was vital proved to be correct. Burnham later said, "If Delise hadn't been there, nothing would have happened." Even so, it was not going to be easy. Before the interview started, Gelb touched on the delicacy of the subject and noted that there would be enormous difficulties involved in publishing such a story, and Serpico, angry and nervous, snapped, "So it's going to be the same old bullshit. You'll let me down like everyone else." But this was quickly forgotten as the editors and reporters listened spellbound during the long interview.

Burnham had prepared a memo on how he was going to treat the story, and now received a go-ahead. It was to be a thoughtfully worked-out, three-part series with much of what Serpico had detailed scattered throughout all three installments, the first one measuring the extent of corruption in the Police Department, the second showing how the corruption had developed, the third suggesting what could be done about it.

After the interview and after he had finished writing the articles, Burnham, like any reporter, became fretful when they were not published at once. But when he went to his editors, he was assured that the series would run, that *The Times* was simply waiting for a hook to hang it on.

More time passed and the story still did not appear. Then, by accident, Burnham got the opportunity to provide the hook he had been told was needed to print his story. On a Saturday in the middle of April a friend took Burnham along to a cocktail party given by Richard Aurelio, who had been Mayor Lindsay's campaign manager in his successful bid for reelection, and was now a deputy mayor in the administration and Lindsay's closest political confidant. Also at the party was Thomas B. Morgan, Lindsay's press secretary. Burnham had not met Morgan before, but seized the chance now to talk to him, and in the course of their conversation Burnham mentioned that *The Times* had a story in the works involving corruption in the city, but did not go into any specifics.

The following Tuesday, April 21, Burnham telephoned Morgan at his office in City Hall, reintroduced himself and reminded him of the story he had spoken to him about at Aurelio's party. Morgan said that he indeed remembered both Burnham and the story he had alluded to, and then Burnham told him it was about police corruption and that it was going to be a blockbuster.

Instantly—"within five minutes," as Morgan later recalled—he was in Mayor Lindsay's office with the news. Deputy Mayor Aurelio was notified, as was Police Commissioner Howard R. Leary, whom Lindsay had brought to New York in early 1966 from Philadelphia, where he had occupied a similar post. A hurried series of conferences followed, and it was finally decided to beat *The Times* to the punch: The Mayor would immediately form and announce a committee to look into allegations of police corruption in the city.

There were some interesting side effects. Serpico got a call from Jay Kriegel, almost three years to the day after he had gone to him to report the corruption he had personally encountered—including the three-hundred-dollar envelope and Captain Foran, and the pad in the 7th Division. Serpico had not seen or talked to Kriegel since.

Kriegel's dialogue on the phone was filled with his usual disjointed sentences, about how busy he was, and, as if to demonstrate this, he continually told Serpico to hold on while he took other calls. But out of the twenty-odd minutes of talk, a theme did emerge in Kriegel's rush of words. "I feel terrible about this whole fucking thing," he said. "Obviously we just didn't have very good communication in the last conversation. We ought to do something again. I know what happened in that last disaster. . . . I'd like to sit down and talk about the problem again [and] see what kind of approach would come to my mind again." Kriegel added that "if that sounds rotten and uninspiring, I'm sorry about it." He said that he had spoken to David Durk, and "I feel like a bastard."

Serpico replied that he didn't see much point in another meeting, but that he was curious about one thing. Had Kriegel relayed what he had told him to Mayor Lindsay? Was Lindsay aware of it? At this point Kriegel grew vague. "Not directly," he said. "We talked about the problem. He had the same problem I did. It had to be referred to the Department of Investigation, the Police Commissioner." They had the "legal authority" to follow through on the allegations.

Then on Thursday, April 23—two days after Morgan reported *The Times* story to Lindsay, two and a half years after Serpico had first told Chief McGovern about Captain Philip Foran and the three-hundred-dollar envelope, months after Serpico raised it again with McGovern and asked what he was doing about it, and with the Police Department's high command now in a frenzy of activity—McGovern finally questioned Foran.

Foran denied practically everything. He conceded that Durk had brought Serpico to see him. He said that he could not recall whether he was actually in the Department of Investigation at the time, although both Serpico and Durk said that the meeting had taken place in August 1966, six months after Foran had taken up his post there, and although Foran did remember that during the meeting Serpico had said he was on "riot duty"—an assignment which Serpico's official record showed took place in the summer of 1966.

Captain Foran flatly denied ever seeing an envelope with or without money, or ever advising Serpico that if he pursued the matter, he would wind up "facedown in the East River," or ever agreeing that Serpico should turn over the envelope to any sergeant. All Serpico did, according to Foran, was to tell him that he expected to receive a payoff to pass on to his fellow plainclothesmen, and when Foran suggested that "we dust the money," Serpico refused because he would be "marked lousy or something like that."

Under questioning by McGovern, Foran admitted that he had made no notes of the meeting with Serpico and Durk, nor did he report it to his superior, Investigation Commissioner Fraiman. Presumably by way of explanation, Foran said that Durk had once recommended Serpico for an assignment in the Department of Investigation, and that he had made the inquiries into Serpico's background and had received confidential information that Serpico was a "psycho." Foran also told McGovern, "I don't know whether I should say this. There was some indication that he had homosexual tendencies. This, of course, is confidential, and I may not indicate my source even if I can remember, which I don't know if I do."

Throughout the meeting with Serpico, Foran said, he was "very, very apprehensive" that Serpico was "trying to set me up." His big mistake, Foran insisted, was not immediately throwing Serpico out of his office, but, "because of my, I guess you might say, fatherly approach to life, I let the man prattle on."

While it was one thing to have a cop like Serpico wandering around alone trying to convince the city to do something about police corruption, getting *The New York Times* into the act was quite another; and the same day that Foran was being questioned— Thursday, April 23—Mayor Lindsay formally announced the creation of a five-man committee "to review all city procedures" for investigating possible corruption in the Police Department. The committee would be headed by the City Corporation Counsel, J. Lee Rankin, and would also include Robert K. Ruskin, who had succeeded Fraiman as Commissioner of Investigation, Manhattan District Attorney Frank S. Hogan, Bronx District Attorney Burton B. Roberts, and Police Commissioner Howard R. Leary.

At *The Times* Burnham rushed excitedly to the metropolitan editor, Arthur Gelb, with Lindsay's announcement, and Gelb agreed that at last the paper had a hook to run Burnham's story. Burnham's original three-part series was torn apart throughout most of Thursday night and rewritten by him and other reporters and editors under Gelb's direction so that all of Serpico's charges, plus other examples of graft that Burnham had gathered on his own, were lumped together in one package.

Around noon the next day A. M. Rosenthal, the managing editor of *The Times*, read the story and approved it. Burnham then called press secretary

Morgan, and told him that the corruption exposé would appear the following morning and that he was sending down two copies "for comment."

Morgan hurried into the Mayor's office with one copy and sent the other to Commissioner Leary, who was holding a meeting in his own office at Police Headquarters, a large, mahogany-paneled room featuring over a mantelpiece the stern features of a predecessor, Theodore Roosevelt. Present was First Deputy Commissioner Walsh, Supervising Assistant Chief Inspector McGovern, and Captain Foran. Foran was in the middle of repeating the denials he had given McGovern when *The Times* story arrived.

Leary and the others barely had time to go over it before the Mayor summoned them. In Lindsay's office they joined another dour gathering that included Deputy Mayor Richard Aurelio, City Corporation Counsel Rankin, Investigation Commissioner Ruskin, Morgan—and, hunched over by a window, chewing on a fingernail, Jay Kriegel.

Parts of the story were still being passed around while the Mayor maintained an icy reserve, his usual manner in front of a group as large as this when he was especially angry. Not everybody there was in accord. Ruskin and Leary, in particular, did not like one another. Ruskin had once asked Leary how he handled corruption, and Leary said, "Walsh handles it," and Ruskin had said, "Well, how do you decide what to do about the complaints you receive?" and Leary replied, "I don't see them until after they've been investigated."

The main thrust of the meeting, which lasted about

an hour and a half, was how to fend off charges in the story. Leary said that it was "a lot of general crap," and blamed "that psycho cop" for it, egged on by his "college pal." This was not exactly the response that Mayor Lindsay had in mind, and finally a statement was worked out which Morgan sent back to *The Times*.

The tenor of Lindsay's statement was that City Hall had everything well in hand: "Police Commissioner Leary has advised me that many of the allegations in this story came from one particular patrolman and were reported to the department in 1967. . . . The department investigated these allegations, referred them to the Bronx district attorney's office and, as a result, a number of indictments were handed down."

Frank Serpico did not have many laughs left in him, but a sentence in the Mayor's statement did produce one: "This government must root out corruption and wrongdoing with every means at its command."

Former Investigation Commissioner Arnold Fraiman, now Justice Fraiman, having been appointed to the State Supreme Court with Lindsay's backing, issued a separate, unctuous declaration that a plainclothesman had furnished him with information about corruption that was "extremely general in nature." Answering a charge in *The Times* story that he had refused to allow a bug to be placed in a 7th Division surveillance truck to overhear cops discussing corruption, Fraiman went on to say that it "would have been a blatant violation of law for the Department of Investigation to do this," neglecting meanwhile to explain why such electronic devices and secret recorders were standard equipment in his department.

The Times ran its story on April 25, 1970, under a front-page headline that said:

GRAFT PAID TO POLICE HERE
SAID TO RUN INTO MILLIONS

The story created a sensation, and for weeks at a stretch police corruption and police shake-ups were page-one topics in *The Times*, the New York *Daily News*, and the *New York Post*, and nightly leads on television and radio broadcasts. To the dismay of the Mayor, whose Presidential ambitions were increasingly evident, the scandal involving his administration became national news as well.

Police Commissioner Leary stirred up more controversy almost at once. While the first story in *The Times* did not identify any sources by name, Leary was perfectly aware that the bulk of it had been supplied by Serpico. Nonetheless, four days later, in charging smear tactics, "McCarthyism all over again," he said that *The Times* had based its report on the word of "prostitutes, narcotics addicts and gamblers, and disgruntled policemen."

As a result, any hope the Mayor entertained that the committee he had appointed with City Corporation Counsel Rankin at its head would smooth things over quickly went by the boards. Five New York City congressmen, led by Edward Koch, who had succeeded Lindsay in the 17th Congressional District, directed their attack at Commissioner Leary's membership on the committee, demanding to know how Leary could investigate himself. At first Rankin defended his group, claiming, "I don't think that an outside committee having no familiarity [with the problem] could do the work with the same skill and effect required of us," and in another interview said he was confident that any statements Leary had made "will have no effect on the action of the committee and its dedication to carry out the assignments that the Mayor had given it."

But the pressure from political enemies, the public, and the press proved to be too much, and City Hall finally ran for cover. In a carefully orchestrated scenario, Corporation Counsel Rankin wrote a letter to the Mayor suggesting that because of the "possibility of conflicts of interest" it might be wise to turn over the investigation to a citizens' group with a full-time professional staff. The Mayor said he would think about it, and after a face-saving interlude he announced the formation of a new, independent commission to be chaired by a Wall Street lawyer named Whitman Knapp.

It remained for Commissioner Leary to reflect precisely the poisonous atmosphere in the Police Department, the system that Frank Serpico had fought against for so long. Almost a month after the story in The Times broke, in a desperate attempt to show that he was doing something, Leary ordered a statement to be read at roll-call lineups in station houses throughout the city. In the statement Leary urged every policeman with knowledge of corruption in the department to come forward and report it. Nowhere did he indicate that any cops who responded would or should be praised or rewarded. All he did was to assure them that they need not fear any "reprisals."

chapter 16

The flood of charges and countercharges following *The Times* story, the disbanding of one investigating committee and formation of another, the denunciations and new accusations, the denials that anything was fundamentally wrong, the promises of action—all this left Serpico essentially unmoved. The attention of the city was at last riveted on the issue of police corruption, but it was nothing more than words so far, and it remained to be seen what actually would be done.

For Serpico, the reality of the moment was that he was going to be the key witness against Robert Stanard. Any possibility that he would not be used as a witness had vanished after two trials of Philip Montalvan, a 7th Division plainclothesman, whose unlisted phone number had been found in the possession of the numbers banker Manuel Ortega. All the testimony against Montalvan had come from policy operators, and the first trial had ended in a hung jury, eleven-to-one for conviction. The one juror who held out was overheard to say that he would "never take the word of people like that against a cop." When Montalvan was retried, he was acquitted.

Serpico had a second meeting with District Attorney Roberts and Chief Cooper over the question of his testifying. It was not much of a battle; after the session the previous December, Serpico knew that he would be a witness, had really known it the moment he had agreed to go before the grand jury a year before. The big fight was over Serpico's beard. Roberts wanted him to shave it off, and Serpico resisted. That was the trouble with the world, he argued, everything was based on appearances, not the truth.

"Look, Frank," Roberts said, "I'm with you, but the fact remains that there may be somebody on that jury who just doesn't like beards, so why take a chance?"

Finally Serpico compromised; he removed the beard but left his mustache, which had grown to handlebar proportions, and balanced it off with a conservative gray flannel suit.

Serpico appeared against Stanard on June 18, 1970. The night before he had spent a restless, even anxious night, unable to sleep, walking through the Village streets with his sheepdog, Alfie, thinking about what had happened over the past three and a half years and what it had come down to, and at last just before dawn dozing off fitfully in a chair for perhaps an hour.

The role he had played in the drama had been an open secret for many months, but there was a finality, a point of no return, about his actually walking into court and directly confronting the man. He wondered how he would feel, seeing him like this face-to-face. But that morning, after he had been sworn in, and the moment had come, when he looked down at Stanard seated at the defendant's table staring back at him, he felt no emotion at all. Suddenly, to Serpico's surprise, Stanard was just another criminal.

Stanard was being tried for perjury in the first degree, for swearing to the grand jury that he had never taken part in a 7th Division pad. Besides Serpico, the chief witnesses against him were Juan Carreras, the policy banker, his wife, Dolores, and the informant Victor Gutierrez. Serpico himself had not been present at the specific incidents the trial would cover, but he was allowed to testify on the basis of a precedent set by a New York State appeals court that permitted a witness to provide background information that gave perspective and relevant "materiality" to a particular case. He could also testify to his conversations with Stanard about the Carrerases, about how they were on the pad, paying off, and to leave them alone.

Serpico quietly recounted his experiences with Stanard as a plainclothesman in the 7th Division, and then Stanard's lawyer began his cross-examination. His initial sally was aimed at Serpico's character. "Were you ever given a summons in a police station?" he asked.

"Yes," Serpico replied. This was a reference to an obscure incident which had occured shortly after

Serpico became a plainclothesman in Brooklyn. Bushy-bearded, his hair long, and dressed in an old shirt and dungarees, he was driving to work one day in his own car, when a red-faced, white-haired traffic cop stopped him for not following his signals while crossing an intersection. Serpico said he had not seen the cop, and the light had been green. The cop made a few sarcastic comments about Serpico's appearance and about hippie types in general. A terrific argument ensued, and by the time Serpico had identified himself as a police officer, it was too late for the traffic cop to back down. The upshot was that, after a great deal more wrangling, Serpico received a suspended fine for not obeying a policeman's instructions.

Stanard's lawyer did not go into these details, content to let mention of the summons speak for itself, apparently hoping to indicate that it had left Serpico prejudiced against all cops. He paused theatrically, then went on. Had Serpico ever been assigned to the Bureau of Criminal Identification?

Yes, Serpico, replied, that was correct.

Did Serpico recall the inspector who had been his commanding officer?

Yes, he did.

"Well, did some incident happen that caused your transfer out of the BCI?" Stanard's lawyer demanded with a leer. "Was there some incident in the men's room down there?"

Oh, boy, Serpico thought, they've really gotten together to do a job on me. The false accusation that he had been consorting with another man in the latrine appeared nowhere on his official record, of course, and he would have been delighted to tell the whole story. But the assistant D.A. prosecuting the case, Robert Koppelman, jumped to his feet, saying, "Oh, I object to this nonsense," and the objection was sustained.

That ended the assault on Serpico's character and reputation, but Stanard's lawyer spent nearly four more hours trying to shake his testimony about the 7th Division pad. When he failed to get anywhere, the lawyer was finally reduced to demanding to know why Serpico had not reported his allegations earlier than the grand jury minutes showed.

This was another story Serpico would have loved to relate in detail, but Koppelman once again was up objecting, and was sustained.

Then Victor Gutierrez and Dolores and Juan Carreras took the stand. Their combined testimony, each filling in parts of the story, established that Juan Carreras had been arrested on several occasions as a South Bronx policy-numbers operator until the early summer of 1967, when Serpico's old partner Carmello Zumatto and another plainclothesman signed him up for the pad. Carreras had to fork over a $2,000 initiation fee. His monthly payoffs afterward totaled $1,650—$150 to the plainclothesmen in the 48th Precinct, where his operation was located, $500 to the 7th Division, $800 to the borough plainclothesmen, and $200 "for the detectives." The $800 to the borough, Carreras said, "did not include lieutenants and up." These payoffs, moreover, covered only himself, his wife, his bodega, and an apartment he used as an office for his numbers business.

Thus protected, the Carrerases were averaging

$7,000 to $10,000 a week in bets until December 23, 1967, when a heavily played number—542—"hit," and Juan Carreras lost more than $20,000. Two weeks later, on January 6, 1968, the same digits in a different sequence—452—hit again, and while his loss this time was not so large, the two together wiped out all the cash reserves the Carrerases had.

Carreras testified that he did not know where to turn. But almost immediately—the next day, in fact—Victor Gutierrez entered the picture and told him that he was acquainted with a major numbers banker in Jersey City named Ricardo Ramos who wanted to expand, and maybe they could make a deal. This was big-league stuff: Ramos was listed by both federal and local law-enforcement agencies as a close associate of an East Coast Mafia power, Joseph (Joe Bayonne) Zicarelli. Carreras agreed, and Gutierrez said that he would produce Ramos on Monday, the 8th.

Then, according to Carreras, he suddenly received a call from William McAuliffe, a plainclothesman attached to the 48th Precinct. Carreras explained that he could not afford his monthly pad payments because of his last loss, and McAuliffe said he had heard about his bad luck, but wanted to see him anyway, and Carreras told him that he was going to have a meeting in the afternoon at which he hoped to straighten everything out.

Four plainclothesmen showed up for the meeting almost simultaneously with the arrival of Gutierrez and the New Jersey numbers banker Ramos. They were identified by Gutierrez and the Carrerases as "Bob" Stanard and "Jim" Paretti from the 7th Division,

McAuliffe and Andrew Taylor, also from the 48th Precinct. At the meeting Stanard urged Ramos to take Carreras on. "Stanard told Ramos that my money was good," Carreras testified. "He vouched for me—that I did good work." Ramos then agreed to back Carreras. He would pay off at six-hundred-to-one odds, whatever the amount, if a number hit, and he would assume responsibility for the pad. Carreras, in turn, could keep thirty percent of the weekly action, out of which he would have to pay his collectors; he could retain an additional five percent if he wanted to "pay the cops" himself, but Carreras replied that he preferred to leave this to Ramos, and noted he was already "a month behind." Ramos told Stanard that he would "pick up" the month the next day, and the meeting was over. It had all been very cut-and-dried, with the police and the underworld matter-of-factly continuing their profitable partnership.

When Stanard took the stand in his own defense, he denied that he had ever discussed a pad with Serpico, denied each of the incidents that Serpico had described involving Stanard and the pad, denied ever receiving payoffs from the Carrerases, denied that he had ever conferred with them about a pad. On June 30, twelve days after his trial began, Stanard was found guilty. He faced a maximum prison term of seven years, was actually sentenced to one-to-three years in a state penitentiary, and was paroled after serving a year.

On July 1, 1970, the day after the Stanard verdict was returned, Bronx D.A. Burton Roberts sent a letter to

Police Commissioner Howard Leary. He wished, Roberts wrote, to "commend to your attention Patrolman Frank Serpico, Shield #19076," and noted that "approximately two or three years ago, the above came forward and exposed a pattern of corruption" regarding gambling laws in the Bronx, and as a result of an investigation by superior officers in the Police Department and the Bronx District Attorney's office, "some eight police officers were indicted."

Roberts pointed out that Serpico had testified not only before the grand jury but also for "the People vs. Robert Stanard, which resulted in the conviction of a corrupt police officer." He added that Serpico had "carried out a duty all too often ignored by other police officers," that "he had exhibited high moral courage," and that "this fine police officer had acted without any thought or expectation of advancement or promotion.

"Perhaps virtue is its own reward," Roberts concluded, but he hoped that Commissioner Leary would "see his way clear" to rewarding Serpico with a detective's shield, adding that this might "encourage others to come forward."

Roberts sent a letter along the same lines to Mayor Lindsay, and waited in vain for an answer from either him or Leary. Finally Lindsay saw fit to post a reply— of sorts. On July 28, in a brief message addressed to Serpico, the Mayor acknowledged that "Bronx District Attorney Burton Roberts has written to me praising your cooperation in the recent investigation and trial conducted by his office. I know that all New Yorkers deeply appreciate the moral courage you have exhibited."

That was all, except for a notation at the bottom of Lindsay's letter indicating that copies of it were being forwarded to Leary and to Roberts.

There was still no answer from the Police Commissioner. Roberts began trying without success to get in touch with him by phone. Working up a fine rage over this high-handed treatment, Roberts persisted and finally got through to Leary, and told him he was calling about the letter concerning Serpico and the detective's shield.

"He's a psycho," Leary snapped.

"In that case, perhaps the department needs more psychos," Roberts replied. "What about the shield?"

"You don't get rewarded for what you're supposed to do," Leary said.

Roberts asked who else in the department was doing what he was supposed to do. "Seriously, I really think he should get it."

There was a pause. "Not while I'm Police Commissioner," Leary said, and that was the end of the conversation.

Following Stanard's conviction, William McAuliffe, the plainclothesman from the 48th Precinct, also went on trial, was convicted, and was sentenced to a year in jail. Zumatto, Paretti, and Taylor were allowed to plead guilty to lesser charges and were given suspended sentences on condition that they resign immediately from the force, forfeiting all rights and benefits.

There were other trials, in court and within the Police Department, some convictions, some dismissals, some cases pending, practically all of them involving lower-echelon men. But essentially, for Frank Serpico, it seemed to be over, and he brooded over how little actually had been accomplished.

Throughout these months he had become close to Chief Cooper, and Cooper in his gruff, blunt way tried to cheer him up. Often their talks turned into shouting matches. Things weren't that bad, Cooper would argue, something had been achieved by the convictions, a man could still do a job in the department, and who knew what changes the Knapp Commission would bring about. Serpico listened. How different his career, his *life*, might have been if he had met men like Delise and Cooper earlier, and had been able to work under them.

Serpico discovered a curious side to Cooper. Whenever Serpico started lambasting the department, the older man would leap to its defense. If Serpico kept his mouth shut, however, Cooper would wind up being just as critical or even more so, about the department, about its leadership, where it was going, what needed to be done.

Cooper not only talked, but he acted, and one particular episode sealed their relationship. Serpico's brother Pasquale, like many merchants in poorer parts of the city, kept his small grocery store open on Sunday—in possible contravention of the state Sabbath law. The law did not flatly forbid certain stores from doing business on Sunday, but it did feature a complicated list of items, some of which could be sold, some not, and others only during specified hours.

The result was a clear invitation for extortion by the police. While the individual instances of graft were not large—two, three, or five dollars—the New York State Joint Legislative Committee estimated that the citywide total of such shakedowns amounted to more than six million dollars annually. And for a store owner like Serpico's brother, even though he was not violating the law—not, for example, selling beer until after the prescribed hour of one P.M.—it was easier "to pay the two dollars" than to spend the time and money fighting a summons in court.

Pasquale first told Serpico that he was being shaken down when Serpico was appearing before the Bronx grand jury in 1968. Serpico reported it at that time to Chief McGovern. "You always come to me with this nickel-and-dime chickenshit," McGovern replied irritably, and likened himself to a general in the army with Serpico as a private who kept running to him about somebody stealing a couple of eggs from the mess hall.

McGovern nonetheless promised to have the matter looked into, but Serpico's brother told him that nothing had changed. So Serpico rode over to Brooklyn on a bicycle one Sunday afternoon with a recording device in his knapsack. He found one of the two cops who were shaking Pasquale down, and enticed him into admitting that they were not only doing it, but would continue to do it.

The cop made it clear that the fact that Pasquale's brother was also a policeman had no bearing on the matter. "Look, I explained that to him today," the cop said. "He says 'OK, I'll go along. . . .' "

"Oh, he did?" Serpico said in a tone he hoped was appropriately innocent.

". . . and he better mean it," the cop went on, "because I was the first one to go in there when he first opened up. I spoke to him and he said he'd go along with it."

"Yeah?"

"So, uh, the next time somebody went in there, he says, 'Yes?' When a cop walks into a store on a Sunday morning, there's only one thing—you don't ask the guy, 'Yes?' You don't embarrass him in front of kids, and you don't talk to him in front of people. You talk to him on the side."

"He's a slow guy," Serpico said, as if his brother was a cross he had to bear through life. "He don't belong in that business."

"I tried to explain to him. I says, 'Basta! Don't be passing money to me in front of kids. If you want to talk to me, come in the back and we can talk like two human beings.' So he said the same thing when I walked in this morning. He says, 'Yes, what can I do for you?' I spoke to him in Italian. I says, 'Look, I'm Italian, you're Italian, we should get along together. There's only one thing you can do. Do what we agreed to do the first week you opened.' "

"Well," Serpico said, "he told me he got another summons today. I'm telling you he's slow, and I just don't want him persecuted because he didn't want to play ball."

"I gave him the summons," the cop said. "He's

going to have to go along with me the same way all the rest of the stores are going to have to go along, or I'm going to give him a summons every day I'm working."

Serpico told McGovern about the tape, but McGovern expressed little interest in it. Then, two years later, right after the conviction of Stanard, Serpico's brother called in a panic and said that he was finally being summoned as a witness in departmental proceedings against the two cops, that it would be just his word against theirs, and what should he do?

Serpico immediately went to Cooper and explained about the recording, and how it had been ignored. Within an hour Cooper dispatched a car to take Serpico to his safety-deposit box to get the tape. Because of it the two cops were found guilty and later dismissed from the force.

Afterward an incensed police officer confronted Serpico. "How could you do that to other cops for a lousy two dollars?"

"Oh, I see," Serpico said, "just because it's only a lousy two dollars that makes it all right."

In one of those ambiguous bureaucratic moves the Police Department seemed so adept at, Serpico was at last transferred to the Detective Division—but still carrying his patrolman's "tin" shield, without being given the gold badge of a detective—and assigned to Narcotics, Brooklyn South. His transfer had come through before the Stanard trial, but because of the preparation required for the case, he did not begin working full time as a narcotics cop until the summer of 1970.

Whatever else his fellow officers knew about him in Brooklyn South, Serpico's reputation as a karate expert and a crack shot had preceded him, and his reception, if not friendly, was at least correct. A sergeant told him that his arrest quota would be four narcotics felonies a month and assigned him to a four-man team. If anything "really big" came up, he was further instructed, it was to be turned over to the Special Investigating Unit of the Narcotics Bureau.

But nothing big ever did, and Serpico quickly found that the police activity against heroin traffic and addiction was as futile and pointless as his nights on the Times Square prostitution detail had been. Theoretically the key targets were the distributors of the drug, the "dealers." But almost all the so-called dealers arrested by the team Serpico was on, and others like it, were actually no more than small-time pushers or addicts passing on an extra supply to other addicts. As far as the arrest quota was concerned, it did not make any difference. Even the sale from one junkie to a friend of a three-dollar "bag" of heroin—the smallest unit being traded at the time—qualified as a felony, and what happened later in court was of little consequence. Because of crowded judicial calendars, a prisoner was usually allowed to plead to possession of the drug, a misdemeanor, and could count on a light sentence; often the sentence was suspended, and if the addict agreed to inform on other addicts, he was simply placed on probation. The rationale many narcotics cops had for these arrests was that at least they "keep the junkies off the street." It reminded Serpico again of how whores were run off the street one night only to turn up on the same old corner the next.

He would have said the hell with it and quit but for his stubborn resolve to see how long it would take to receive the detective rating he had always wanted and, by every measurement, was now entitled to. There was, besides, another factor. In September, slightly more than four months after the police corruption story appeared in *The Times*, Police Commissioner Leary abruptly resigned, both he and Mayor Lindsay vehemently denying that this had anything to do with the forthcoming Knapp Commission investigation of the department, and the Mayor replaced Leary with Patrick V. Murphy, a former head of the New York Police Academy who was then running the Detroit force, and who enjoyed a rare police reputation—that of a "law-and-order liberal." In short order, the department began to undergo a traumatic upheaval and reorganization—with anticorruption efforts and better law enforcement supposedly the battle cry of the day—and Cooper and Delise urged Serpico to stick it out, to see what would happen, and to recognize that the department needed officers like him.

Heroin, meanwhile, continued to pour into the city without letup, accompanied by all the insane acts committed by addicts desperate for a fix; and Serpico became so dispirited at the meaningless steps being taken against it that he let his partners take all the credit for the arrests the team made. He was sure that the same standards applied to narcotics as to illegal gambling. If the cops wanted to, they could eliminate a great deal of the narcotics business almost overnight. But there was little Serpico could do on his own. In Brooklyn South, as he watched and waited for promises of change in the department to come through, he was now a marked man, purposely kept isolated from what was truly going on. He might as well have been living in a cocoon. Even in the company of his partners, with whom he worked daily, there was a constant, subtle tension of mistrust.

Then Serpico got a glimpse of what he suspected was happening all around him. He was walking along the street when a narcotics detective pulled up beside him in a car. "Hey, you're Serpico, right?" the detective said.

"Yeah, what's up?"

"Get in, I want to talk to you." Serpico slid into the seat next to him.

"You're not wired, you fuck, are you?" the detective said.

Serpico shrugged and raised his arms, let the detective pat him down, curious as to what he would say. Satisfied that he had nothing to fear, the detective said, "Look, I can understand that stuff you did up in the Bronx. That was just chicken feed. How'd you like to make some real money?"

"Hey," Serpico replied, "maybe I should be doing the checking. What is this, a setup?"

"Come on," the detective said, "I'm trying to do you a favor."

"How?"

"I'm telling you. We don't fuck around with a lousy

eight hundred a month. Last week, like, there was this distributor, and he had three guys making pickups, and each pickup was for forty Gs, and we just waited until he got all the money, and we hit him. You know what that is, four ways? Now I'm not saying you make a score like that every day, but it's there. You never know."

Serpico wondered what had happened to the tortured distinctions he used to hear about clean money and dirty money. "How come you're telling me all this?" he asked.

"Well, you know, you got a reputation. You make a lot of people nervous."

"So what am I supposed to do?"

"Wise up," the detective said.

The approach, complete with its inherent threat, left Serpico as enraged as he had been at the meeting in the Bronx district attorney's office when he learned how little the 7th Division investigation was destined to accomplish. First from one side, then from the other, the message was the same: be reasonable, go along. Well, he would not. He was being pushed into a corner again, and this time it was infinitely worse—not gambling payoffs but heroin. Serpico was determined to push back.

But he never got the chance.

chapter 17

On Wednesday, February 3, 1971, Frank Serpico spent most of the morning cleaning up his apartment, buying some groceries, and taking a pile of dirty clothes to a neighborhood Laundromat.

While he and his partners were supposed to put in a regular eight-hour day, in practice they, like other narcotics teams, usually worked only when a specific arrest or surveillance was at hand, and kept in touch with each other through a message center at the Narcotics Bureau headquarters in Manhattan. The day before, the team leader—Gary Roteman, a lean, graying man four years older than Serpico—had said that there wasn't anything on for Wednesday, except possibly in the evening, but he would have to check first with one of his informants; and Serpico had replied that he would be at his apartment in the late afternoon if he was needed.

So after he had finished his chores, Serpico went, as he normally did once a month, to the firing range at the Police Academy, and practiced with his Browning automatic, his .38-caliber service revolver, and the snub-nosed .38.

When Serpico was back in his apartment, Roteman called to say that his informant had something for them, there might be a couple of arrests involved, and they better try for them, since it was early in the month and they had not done too well in January. Serpico customarily carried his Browning and his snub-nosed .38, and kept his service revolver in a bank vault so nobody could get at it in case his apartment was broken into while he was away, but by the time he had returned from the firing range the bank was closed, and he brought all three guns with him. When he walked into the Brooklyn South Narcotics office in the 94th Precinct and took off his jacket, one of the men said, "Jesus, what are you expecting, a revolution?"

Serpico went with his partners to a luncheonette for coffee and pie. Besides Roteman there was Arthur Cesare, darkly Italian with black hair and a straight nose, and a newcomer, Paul Halley, who had a pale, round face with tiny eyes which made Serpico think of two raisins in a cookie. One topic was notably absent in their conversations with him. None of them had ever mentioned, even obliquely, anything about the 7th Division pad.

Cesare was saying, "You know what, my kid came home from school, and you know what they're teaching him to sing? That fucking 'We Shall Overcome.' So I told him, 'I don't want you singing that, you tell the teacher.' It's all because of that fucking Commie, Martin Luther King."

"Oh, I didn't know Martin Luther King was a Communist," Serpico said in mock astonishment.

"Sure, he was a card-carrying member."

"Gee," Serpico continued, "that's good information. Where'd you get that, from one of your informants?"

"Come on, everybody knows it."

"Well," Serpico said, no longer trying to disguise his sarcasm, "you took care of it. You told the kid not to sing. That's nice. What's he going to do when he goes back to school? You going to be standing behind him when the other kids ask him why he isn't singing with them?"

Cesare glared at Serpico. "Don't worry," he said. "I'll talk to that teacher myself."

"Let's go," Roteman said.

Roteman's informant was a young Puerto Rican whose most distinguishing feature was a pair of gold-framed glasses. Serpico didn't even know his name, although Roteman had used him once or twice before. He had been an addict when he was turned into an informant, and was supposedly off heroin now. He was probably informing, Serpico thought, and dealing on the side—it usually worked that way. But as Roteman said, "He gives good information," and that was all that counted.

When they got to the informant's apartment house, Halley stayed in the car, and Serpico decided to go up with Roteman and Cesare. Bored with the prospect of another meaningless heroin arrest to meet their monthly quota, the idea of sitting in the car making small talk with Halley only depressed him more.

The informant let them in, and said his wife was out and he had to wait for her to come back to take care of the children. One was a boy, perhaps five, the other a girl a year or so older, both with great black eyes, and Serpico talked to them in Spanish while Roteman and Cesare huddled with the informant in a corner of the room. The children were fascinated by Serpico's beard, and he let them tug at it. Christ, he thought, they're the only ones here I can relate to. Finally the informant's wife arrived. She was an attractive woman, and she excused herself for being late and anxiously asked if anyone wanted coffee. His two partners said no, they had to go, but Serpico thought that if she was gracious enough to make the offer, he would accept. Besides, he liked the strong coffee Puerto Ricans drank. So everyone stood around while he sipped the coffee mixed with hot milk and morosely wondered if the informant's wife knew what her husband was doing.

Then they all left. In the car the informant directed them down Driggs Avenue, a north-south street in the Williamsburgh section of Brooklyn. In the last block before it terminated at the approach to the Williamsburgh Bridge to Manhattan, they passed a fairly large five-story building on the right, and the informant said that he knew there was a man in there—a man named Mambo—who was a pusher. They circled two blocks and came back down Driggs Avenue, and stopped on the left next to a deserted school yard, just short of an intersection, a little over half a block from the building.

The plan was for the informant to station himself at the building's entrance. If anybody went in and came out, he would try to determine if a "buy" had been made. He knew most of the people in the area and could talk to them, and when he discovered someone who was "dirty"—who had heroin—he would signal it by removing his glasses and wiping them with a handkerchief.

It was dark and Driggs Avenue was dimly lit, but there was a street lamp near the building entrance that cast enough light for them to watch the informant with binoculars. Serpico's three partners were gossiping so much, however, that they often forgot to look through the binoculars when their turn to use them came, and Serpico, anxious to get this over with, finally took them and kept them. He was in the back of the car, and he rested the binoculars on the top of the front seat and peered steadily through them, growing more irritated every minute. It was now past nine P.M. and they had been there for more than an hour. The informant maddeningly disappeared inside the building from time to time, and Serpico strained his eyes in the darkness waiting for him to show up again. Cesare, in the middle of telling a joke, bumped against the binoculars, and Serpico exploded, "Are you guys working, or what?" Occasionally people walked by the car, and some of them looked incuriously, and

Serpico said hopefully, "If we stay here any longer, we're going to get spotted," and Roteman said, "Relax, nobody's going to spot anything."

Suddenly Serpico saw the informant take off his glasses and wipe them. "Hey," he said, "he just signaled! It must be the woman with the shopping bag." The woman had come out of the building and turned right, away from them, down Driggs Avenue. When she turned the corner, they followed in the car and pulled alongside of her halfway up the side street. Serpico's partners jumped out of the car while he remained inside. The three men surrounded her, and one of them was looking in the shopping bag. It was so dark on the side street, that Cesare came back to the car for a flashlight. One did not work, and when he found a second one under the seat that did, its batteries were almost gone. "Christ, why doesn't somebody check these things?" he muttered angrily. From the car Serpico could see that neither the shopping bag nor her purse had produced anything, and he knew his partners were in for trouble. The woman, in her early twenties, was beginning to protest loudly after her initial surprise, and to search her more thoroughly could lead to all sorts of complications. According to the regulations, she would have to be brought back to a station house so a policewoman could do it. Finally they gave up and let her go.

When they returned to the car, Roteman said to Serpico, "You were a big help."

"What's the matter? Weren't three of you enough for one woman?"

"Well," Roteman said, "it wasn't a good collar anyway. Let's go back to the same place."

They parked by the school yard again, and when another ten minutes passed with no signal from the informant, Serpico said, "What the hell are we going to do, stay here all night?"

Then Roteman said, "Why don't you go up there and see what you can find out? They won't 'make' you."

It was, Serpico thought, the same old story. The area was basically a Puerto Rican ghetto, and the way his partners were dressed and looked, a ten-year-old kid would instantly recognize them as cops. But Serpico would have no such problem. On this cold February night he was wearing calf-high boots and heavy socks under his dungarees, and a thick woolen turtleneck sweater and a leather vest beneath his padded army jacket, along with a scarf around his neck. He got out of the car. He went into a bodega, bought a can of beer and started drinking it while he sauntered toward the building as if he were just another street bum. The address was 778 Driggs Avenue. In another time the building must have been the neighborhood showplace. It had an ornate, stone-sculptured front and even a name, also in stone, NOVELTY COURT, but now its shabby entranceway was filled with garbage, the paint on the walls peeling off in great chunks.

The informant was standing inside the entranceway, and as Serpico went by, he looked at him and raised his eyes slightly, indicating that he was going to the roof. Two teenagers sitting on a step leading to the lobby glanced briefly at Serpico, and resumed their conversation.

Serpico went up the first flight of stairs. His .38 service revolver was inside the faded gas-mask bag slung over his left shoulder, and the Browning automatic was in its usual position, in a belt holster on his left side, the butt forward. He took his snub-nosed .38 out of the right side of his belt and put it in his right jacket pocket, keeping his hand in the pocket as well. It was a precaution he always followed in any narcotics action when he was alone like this. Gunplay, with policy-numbers operators was practically unheard of, but heroin was different. Junkies were dangerously unpredictable, and dealers who were caught and couldn't buy their way out or otherwise make judicial arrangements knew they faced long prison terms.

He continued to mount the stairs, and admired the intricate designs of the tile landings, now barely visible through the filth. The landings were filled with the smell of beef bones being cooked down into stock, and the lingering odor of fried pork and *achiote*, a reddish spice used to color rice. He heard nothing besides music being played behind closed doors, and the sound of television sets. The stairs were deserted, except for a bedraggled dog stretched out on them just before the roof. He stepped gingerly over the dog, hoping he wouldn't start barking. But the dog did not move a muscle, and when Serpico went out on the roof, he realized why. Probably the most common experience in the dog's life was being stepped over.

A stench of urine hung in the air despite the cold,

and the roof was dotted with lumps of dried feces. In the light cast by a naked bulb at the top of the stairs, Serpico saw dozens of empty glassine bags and, even worse in a way, the flattened tubes of airplane glue that children sniffed before they graduated to main-lining. Here it all was, all the mute horror and degradation of lives made hopeless by heroin, and he thought again of the futility, the hypocrisy, of going after users and penny-ante pushers while the big-time dealers seemed to remain miraculousy untouched.

Off to his right he could see the pretty, bluish beads of light marking the span of the Williamsburgh Bridge, and the spectacular Wall Street skyline across the river, so close that Serpico felt he could reach out and touch it, glittering as though the ghetto rooftop he was on did not exist. After a few minutes the informant joined him. "What happened?" the informant said. "Did you get the girl? She was dirty."

"No, they stopped her, but they weren't sure, and they let her go."

"She was dirty. She had it on her. They should have taken her."

"Well, they didn't," Serpico said, "so that's that. Did those kids downstairs say anything about me?"

"No, they only said you were a freaked-out hippie coming to make a buy."

Serpico laughed, and asked the informant if he knew what apartment the pusher, Mambo, was operating in, and the informant said it was 3-G, on the third floor next to the stairs.

"OK," Serpico said. "You go back down and signal them if anybody makes a buy. I'll stay up here for a minute and see if I can spot something." The building was U-shaped with the two wings paralleling the street, and Serpico thought he might be able to see into apartment 3-G from the roof of the opposite wing. But then he realized that the apartment was situated so that its windows opened on the rear and far side of the building, where he had no vantage point. He picked his way through the mess on the roof and looked down on Driggs Avenue. He could see the informant standing directly below him, and two men on the sidewalk across the avenue.

Just then the two men abruptly crossed over toward the building. Serpico edged back down the stairs, and stopped between the fourth and third floors. There was a faint light on the third-floor landing, but the stairs themselves were dark. He heard footsteps coming up, and through the banister railing he saw the two men stop and knock on the door of 3-G. There was a muffled response, and one of the men asked for Mambo. The door opened slightly, there was more dialogue, and the man who had knocked handed over some money. The door closed and opened again, and Serpico saw something handed out, and the two men started hurrying down the stairs. Serpico followed them into the street. They were headed toward the car his partners were in, and he pointed at the two men, really youths barely out of their teens. His partners jumped out, grabbed them, and found two bags of heroin on the one Serpico had seen passing money into the apartment.

Serpico described what had happened, and Roteman said, "OK, let's get this Mambo."

"How are you going to work it?" Serpico asked.

"Well, you know, you make the buy. You just get the door open for us. What the hell, you speak Spanish."

This was not the kind of assignment Serpico was supposed to undertake. Making a buy was the job of undercover cops in the Narcotics Bureau. But he agreed. In five minutes it would all be over, and he could go home.

The youth with the two bags of heroin was handcuffed and left in the car with Halley, and Roteman and Cesare accompanied Serpico back into 778 Driggs Avenue. The two teenagers, who had ignored Serpico before, were still sitting on the steps in the entrance, and one of them now said to him, "Hi, officer," and he thought, Well, the kid is getting smarter all the time.

As they climbed the steps Serpico asked Roteman, "Did you get that guy's name so I can say who sent me?" and Roteman said, "Oh, I don't know. I think it was Joe or Tony or something—say Joe." The name of the prisoner in Halley's custody was, as it turned out, Luis.

The third-floor landing, like the others in the building, was long and fairly narrow, with two apartments at each end, and more of them along the side facing the stairwell. Apartment 3-G was immediately to the right of where the stairs came up to the landing. Cesare stayed on the stairs, a step or two down, and flattened himself against the wall. Roteman, on the landing, did the same against the apartment door adjoining 3-G.

The door to 3-G was painted red, part of it chipped so that the tin sheeting underneath showed through, and it had a peephole. Serpico put his right hand in his army jacket and gripped his off-duty .38. Then he knocked on the door with his free hand.

He heard someone coming. The peephole opened, and he could see a distorted eye staring at him. "I need something," Serpico said in Spanish, mumbling, "Joe sent me."

The voice on the other side of the door called, "Mambo, hey, Mambo," and the door opened slightly, and Serpico saw that it had a chain lock. He quickly stepped back, lowered his left shoulder, and lunged forward, hitting the door as hard as he could— expecting Roteman and Cesare to be right behind him.

The chain snapped. The door started to give way, and Serpico continued his charge, twisting around as he went, so that his head and the right upper side of his body wound up inside the door. He could see that there was a small alcove on the other side of the door, not much wider than the door itself. The alcove was dark, except for a little light coming from the adjacent kitchen.

All at once the door was being pushed shut again, and in the darkness he saw a figure, possibly two, bent low, straining against it. Serpico yelled, "Police! Hands up!" Whoever was shoving the door back against him had the advantage of being able to use the alcove wall for leverage. Serpico heard someone screaming, "Mambo! Mambo!" and he saw a young, thin-faced man hurry out of the kitchen toward him.

Serpico's body was half in and half out of the

apartment, pinned by the door, his feet wedged awkwardly in the opening. His right arm was jammed against the alcove wall, and he desperately tried to get his off-duty .38 out of his jacket pocket.

He swiveled his head around and looked back through the opening, and saw his partners standing behind him on the landing. They seemed transfixed. "What the fuck are you waiting for?" Serpico shouted. "Give me a hand—*push*!"

He turned back into the apartment, and somehow, in a last frantic effort, he managed to move his right arm enough to get the .38 out of his pocket. Then he saw the pistol pointing at his face, possibly eighteen inches away.

He remembered that it was like watching a film that had been speeded up. He saw the gun, and suddenly there was this enormous flash, not the little flash one usually sees when a gun is fired, but a great, huge flash exploding in front of his eyes, full of colors, all red and orange and yellow, and he felt the heat flood through his head, and he knew that he had been shot. He had boxed a lot as a teenager, and the pain he experienced was exactly that of taking a hard punch on the nose, and he thought, Is that all there is?

Everything seemed to go into slow motion, a dream sequence. Serpico had just brought his own revolver up when the blinding flash went off. The .38 was on single action, the equivalent of a hair trigger, so that even the slightest reflex pressure on it would cause the revolver to go off, and almost simultaneously he fired back.

The sound of the first shot, the one that sent the

bullet into Serpico's head, had been as enormous, as stunning, as the flash accompanying it, as if his head had been inside a barrel and somebody had thrown a cherry bomb in it, and now his answering shot came to him like a faint, distant echo of the first one.

Serpico fell back heavily through the door opening onto the tile floor of the landing, cutting and bruising the right side of his forehead. His feet were still wedged between the door and the doorjamb. He tried to free them, and then he felt himself being dragged away from the door.

He was on his right side, and he could see his blood geysering past his eyes, splattering against the wall inches away, and streaming down it. Cesare was bending over him, and Serpico said, "For God's sake, try to stop the bleeding. Use my scarf."

He saw Cesare start to daub at the bleeding, and Serpico wanted to yell, "Not that way, you bastard," but he was choking on the blood in his mouth and throat, and he could not get the words out.

Somewhere behind him he heard Roteman shouting, "Police! Throw your guns out, and come out with your hands up," and the sound of a woman sobbing hysterically in the apartment, and Serpico thought, I'm lying here bleeding to death and he's telling them to come out with their hands up.

A great weariness seized Serpico. He felt as he did when he was a boy nodding over his homework, so tired, wanting to go to sleep, and now he fought to keep his eyes open, thinking that if he could only keep from losing consciousness, somehow he would not die.

The first words of reassurance he received that he would be all right, that help was on the way, came not from his partners, but from a wizened, elderly man who suddenly knelt at his side. "Don't worry," the old man said to him in a Spanish accent, "I call the police. The police come. Don't worry." And he squeezed Serpico's hand.

Serpico heard the sound of a siren, a police siren, and as he lay there on the filthy tile landing, he heard footsteps pounding up the stairs, and he saw two uniformed police officers, and one of his partners said, "We're Narcs, he's on the job," and one of the uniformed men stared down incredulously at the bearded Serpico and exclaimed, "He's a cop?"

Then, recovering from their astonishment, the two officers hurriedly picked him up, his arms flopping, and carried him down the stairs and out of 778 Driggs Avenue, past the two teenagers and the informant still in the entranceway, and placed him gently in the rear seat of the radio car, turned on the siren, and started racing to Greenpoint Hospital, where Frank Serpico's inert, blood-soaked form was wheeled into the emergency room, and the first nurse to reach him stifled a cry that he was dead when she heard him trying to say something.

While Serpico lay stretched out on the landing outside apartment 3-G, Edgar Echevaria, alias Mambo, the twenty-four-year-old pusher and addict who had gunned him down, made his escape.

The apartment was occupied by one Ruben Tarrats. He was in the apartment that night, as were his brother and sister, who was Mambo's girlfriend. The sister said that when the struggle and shooting began at the front door, she was in another room watching television, and that Mambo, holding up a bloody hand, ran into the room, crying, "It's the police, I'm shot, I'm shot," and then he and her brother Ruben had gone out the window and down the rear fire escape.

Questioning her and the other Tarrats brother, the police learned that among the possible places Mambo might be hiding was the apartment of another female friend in Williamsburgh. The house was staked out, and at approximately two-fifteen A.M. on the morning of February 3, as Mambo was trying to sneak out of a rear window of the second apartment, he was spotted by Patrolman Maxwell Katz and shot in the stomach. Katz had been at home in North Massapequa, Long Island, when he heard a television news report about the shooting, and he had driven into Brooklyn to help search for Serpico's assailant. According to Katz, he saw a man climbing through the window, his right hand wrapped in a handkerchief, his left hand holding a gun. "He went to level the gun," Katz said, "and I fired." In Mambo's possession was the .22 target pistol he had used to shoot Serpico, as well as Serpico's snub-nosed .38, which he had picked up from the floor of the alcove where Serpico had dropped it; one round had been fired.

When the initial fear of the Police Department's high command, that Serpico had been shot by another cop, turned out to be unfounded, there was a collective sigh of relief. But because it was Frank Serpico who had been shot, an inquiry was immediately launched to determine the circumstances of the shooting, whether Serpico had been the victim of a plot.

Both Arthur Cesare and Gary Roteman were questioned in the early hours of February 4, Cesare at 3:43 A.M. and Roteman at 4:07 A.M. Both were told that they had the right to have counsel present, and both waived the right.

Cesare said that just before Serpico knocked on the door he had positioned himself on the stairs to Serpico's right and that Roteman was standing against the door of the adjoining apartment on Serpico's left. He confirmed that Serpico had slammed into the door, but his description of Serpico pinned half in and half out of it, frantically trying to extricate himself, left something to the imagination. "Frank couldn't get the door all the way open. So he asked Gary and I to give him a hand."

The two men were going to Serpico's assistance, according to Cesare, when "a shot came from the apartment, striking Frank." Serpico began "to stumble down." Cesare tried to support him. "As I went down," he said, "another shot was fired. Then I tried to grab Frank, but his foot was wedged in the door [and] as I was pulling him out I saw a gun start to come out of the apartment door. That's when I fired a shot at the gun. That's when Gary said, 'Throw out your guns and come out with your hands up.'"

At the conclusion of his interrogation, Cesare was asked if Serpico had fired his revolver. He replied, "I don't believe so."

Questioned separately, Roteman said, "The door was opened partway. There was a security chain on the door." There was an attempt to "push the door shut." Then, he went on, "I heard a shot. At this time I am directly behind Serpico pushing into him to open the door." Roteman said that he saw the flash of a gun fired from inside the apartment, and that he fired "two shots" in return. He said that he told Cesare "to go and get an ambulance"—which Cesare did not mention in his initial version of what had happened. Roteman added that he went into another apartment in the building and also phoned for an ambulance, identifying himself as a cop and explaining that another cop had been shot. With Cesare outside on the same mission, this would have left the critically wounded Serpico completely unattended.

There was no reference at all to the old man who had knelt next to Serpico as he lay bleeding on the landing and told him not to worry, that he had called the police. But the signal that was broadcast sending the radio car to Serpico's aid was a "ten-ten," which means "shots fired"—the usual code for a shooting between unknown parties—and not the signal "ten-thirteen" that means "assist patrolman" and automatically goes out when it is known that a cop has been shot or is in trouble.

Both Roteman and Cesare insisted that it was they, not Serpico, who finally broke the chain lock; and later in the morning, using the information gained in the interrogations, a police spokesman told reporters that Serpico had knocked on the door of apartment 3-G and tried to make a heroin buy, that the door was partly open, held by a chain, that Serpico attempted to enter, that two shots were reportedly fired from inside the apartment and that one of them hit Serpico, who "fell through the narrow door opening."

How this could have occurred was never explained. Most chain locks, like those in the apartments at 778 Driggs Avenue, average five to six inches in length, just enough for someone inside to see who is at the door—or for passing money in and bags of heroin out. An interim police report dated February 9, 1971, also referred to Serpico's falling "through the narrow door opening," but mention of this was eventually dropped. As an assistant district attorney in Brooklyn noted dryly, "That chain must have been a yard long."

Questioned again on March 19, Cesare recalled, "I fired one shot into the apartment as two of the perpetrators were going out the back window down the fire escape. I then ran down the stairs to make a notification for an ambulance and assistance."

Roteman, on the other hand, now said that it was only after "help arrived" that "we ordered the occupants to come out with their hands up." According to Roteman, one of the shots fired from the apartment whizzed right by him on the landing, and in describing his emotions at the time, he said, "We were in a state of shock, so to speak." He also said, "Myself and Cesare found Frank Serpico to be a willing and diligent officer and [we] will feel his loss since it seems that he will be unavailable for a long period of time."

There was a third hearing in which Roteman and Cesare appeared before an Honors Board for recognition in having helped save Serpico's life. The board decided, while considering the matter, that the discrepancies in their stories were due to "natural confusion," and eventually gave them decorations of Exceptional Merit, the fifth in rank of eight possible awards that the Police Department could bestow. Patrolman Maxwell Katz, who had voluntarily come into the city and wound up capturing Serpico's would-be killer, received the same citation. When Serpico finally met Katz, he commiserated with him. "If it had been anyone else besides me," he said, "you would have been promoted."

The rumors, the whispers, that Frank Serpico had been set up, even without the knowledge of his partners, would persist, and in the minds of many people would never really be resolved. For Serpico himself there was only the bitter memory of the night of February 3; he would never forgive his partners for what he believed was their failure to back him up when he tried to shoulder his way into the Driggs Avenue apartment.

Serpico had been in Brooklyn Jewish Hospital about a week when Arthur Cesare came to see him. The left side of Serpico's face was still paralyzed and puffy, a mottled bluish-yellow. His left eye was black and almost shut. His speech remained slurred from the impact of the bullet fragmenting on his jawbone. He stared at Cesare with his good right eye and whispered, "Why don't you go back to the academy and learn how to shoot?"

"Fuck you, I won't give you this," Cesare said, nervously holding out an envelope. "It's your watch."

"Leave it and get out."

"I found it on the floor."

"Maybe you should get a medal for that."

"What do you know? You were unconscious. You don't know what happened."

"I remember very well," Serpico said.

"I caught you in my arms after you got shot. I laid you down nice and easy."

"Bullshit. How did I get this then?" Serpico said, pointing to the bruise on the right side of his forehead that he received when he fell on the tile landing. "Do me a favor. Get the fuck out of here before I call the nurse."

Cesare hesitated, put down the envelope, and left.

All told, Serpico was in the hospital six weeks. Once emergency surgery to remove the bullet fragments had been ruled out the night he was gunned down, the chief concern to his doctors was that they might have to go into his head to repair the rip in the cerebral membrane. But the massive infusion of antibiotics that they hoped would make this unnecessary did the job, and by the end of February—three and a half weeks after Serpico had been carried out of the building on Driggs Avenue—the rip had mended itself and the sinister, blood-streaked cerebral spinal fluid stopped oozing from his ear.

With the danger of a fatal meningitis infection over, Serpico's neurosurgeons, Aaron Berman and Zeki Ugar, turned their attention back to the bullet fragments, especially the one so close to the carotid artery in his neck. A series of three-dimensional X rays taken while he was still being given antibiotics showed that the fragment had apparently remained stationary. And now, on March 1, another study was performed to determine the precise placement of the bullet fragment next to the carotid artery. A needle was stuck into the artery, dye injected, and then a new series of X rays taken. It confirmed earlier examinations that the fragment was in a lateral position precisely half a centimeter—less than two-tenths of an inch—from the artery, and that it had not moved since Serpico had been shot.

What had to be considered was the threat of aneurysm, or ballooning of the artery wall around the fragment, which could immediately affect the brain or heart, or cause a rupture in the artery itself. But again the decision of Drs. Berman and Ugar was against surgery. Their reasons were threefold: first, Serpico's facial nerves, which were recovering in surprisingly good fashion after their initial paralysis, might be severely damaged; second, in an operation as delicate as this would be, harm might come to the artery or a branch of it; and, third, there was always the chance that they might not be able to find the fragment. Both Berman and Ugar were the sort of neurosurgeons who believed in not operating if at all possible, and leaving the fragment where it was seemed to offer the least risk while everything was going so well. There was, of course, one imponderable. The fragment could start moving at any time, and for the rest of his life Serpico would have to have it periodically checked.

Serpico probably would have been discharged the following week, but he suddenly developed a fever and complained of pain in his left leg. This turned out to be phlebitis, an inflammation of the veins brought on by his being bedridden for so long, and the danger was that it could send deadly blood clots into his lungs. In five days, however, his temperature had returned to normal, and the swelling in his leg was under control.

Serpico spent another week in the hospital. His hospital chart noted that his mood was cheerful, and Dr. Ugar, who had taken such a personal interest in him, came into his room and finally told Serpico the one thing that he had been holding back about his condition. Because of the concussive effect of the bullet against his ear bone he would remain permanently deaf in his left ear. After the first shock of this news had passed, Serpico grew philosophical about it. In view of all other things that might have happened to him, it seemed a small price to pay.

He left the hospital on March 15 in an unmarked police car. He was taken to the headquarters of the Brooklyn borough command, where he collected his belongings, his Browning automatic and his .38 service revolver—the off-duty .38 being held for evidence against Mambo—and a cardboard box containing his clothes. When he got home to his Perry Street apartment and opened the box, he found that the blood-soaked clothes were wrapped in a plastic bag. He tore open the bag to take them out; they were still sticky from the night of February 3. He kept his boots and threw out everything else.

The day after Serpico returned home his leg started to bother him again, and he had to use a wheelchair for about three weeks. Then, when the leg improved, he would limp down to a Hudson River pier, leaning on the sword cane he had bought to use for support—and for the added protection he felt he needed, whatever the circumstances behind his being gunned down—and he would sit in the sun for hours at a time, his sheepdog, Alfie, on watch beside him, trying to keep his mind as blank as possible, trying to forget the ugliness of the past, not even wanting to contemplate the future at this point, content just to watch the river traffic and wait for his strength to return.

Sometimes he slipped into weird reveries, and once he dreamed about giving testimony in the classic way cops were supposed to speak on the witness stand. Serpico often used to parody this stilted language for his friends. "Yes, there did come a time when I observed the perpetrator perusing said slip of paper and making notations on same with a yellow pencil." Everyone would laugh, and Serpico would continue, "The defendant was approached by several unknown males who, after a short conversation, did hand him U.S. currency in bill and/or coin form, which he did accept and place in his right trouser pocket. After the last such transaction I did leave the point of observation where I had been secreted for sixteen hours and approached said defendant and placed him under arrest, whereupon he said, 'Officer, can't we work this out?' "

There, on the pier, as he sat half asleep in the sun, everything seemed mixed up. He was on the stand, but the courtroom was filled with laughing friends, and he was testifying against a prostitute and saying, "I did observe the defendant engage several unknown males in conversation, at which time I did pass by the defendant and say, 'Good evening,' and said defendant did ask me if I was interested in having a good time, to which I did reply, 'I wouldn't mind having dinner and a few drinks,' and the defendant then stated she didn't have all night, that she had a business to run"—and suddenly he was shouting over and over again to his friends in the courtroom, "What's so funny? It's not funny anymore," and woke up, bathed in sweat.

In May, still recuperating, beset by blinding headaches that came down on him without warning, Serpico gave in to the arguments of Chief Cooper and went to Police Headquarters to receive his detective's gold shield from the new Police Commissioner, Patrick Murphy. "What the hell, Frank," Cooper had said, "don't throw it away. You earned it, you deserve it."

It should have been the greatest day of his life, the culmination of everything he had yearned for from the moment he had entered the Police Academy. Now he had the shield, and he didn't care.

Serpico's neighbors in Greenwich Village by this time knew, of course, who he was, and to his astonishment, instead of resenting his being a cop, they started coming to him in droves with problems they had been afraid to go to the police with, or believed the police would do nothing about—a homosexual terrorized by a man he had taken into his home, an elderly shopkeeper being shaken down by a building inspector, a

hippie store that had been burglarized, a white girl trying to get away from a black man she had been living
with who threatened mayhem if she left. The irony
was exquisite. Now he was being called upon to do
all the things he had envisioned to be the proper task
of all police officers—helping people.

Then one night a group of friends was at Serpico's
apartment, drinking wine and chatting, when somebody suggested they form a local vigilante group—the
Village Citizens Committee—and with cheers and
toasts Serpico was immediately elected the head of it.
The committee's greatest triumph—it would be forever known as the Hudson Street Cobblestone Affair—soon followed. Hudson Street, which ran near
Serpico's apartment, was one of the last thoroughfares
in the city paved with nineteenth-century cobblestones. A section of the street had been ripped up to
install new telephone cables, and a girl who was a
staunch member of the committee told Serpico that
people were stealing the cobblestones for souvenirs
and something had to be done.

It really wasn't something that he could get that
worked up about, but the next evening he, the girl,
and another couple were coming out of a delicatessen,
and Serpico said, "Well, speak of the devil." Down the
block, next to a big pile of the cobblestones by an ex-
cavation, a burly man dressed in a white shirt and tie
and a dark-blue suit was picking them up two at a
time and putting them in a station wagon. The girl,
rather frail and small, promptly marched up to the
man and said, "Excuse me, sir, but that's not a very
nice thing you're doing. This area of the Village is a

historical landmark. How do you think the street's
going to look when they pave it again and there aren't
enough cobblestones?"

The burly man looked at her. "Listen, sister, you do
your thing, and I'll do mine."

"Suppose my thing is to have you arrested?" the girl
said, and angrily marched back to Serpico, hands on
her hips. "Well, Frank?"

Serpico sighed and walked over to the man. "Hey,
why don't you do what the lady said?" he asked. "Put
the bricks back and be on your way." Then he took
out his shield.

The man stared at Serpico, at the beard, the flared
denims, and sandals, and at the shield, and replied,
"Shit, I got one of those in my pocket too, and if I
looked like *you* I wouldn't go around showing it."

Serpico was speechless for a moment, and then in
the best official manner he could muster he said, "If
you have one in your pocket, you'll understand. The
citizens of this community have registered a complaint with me, and I'm honoring that complaint. Are
you going to knock it off, or what?"

"OK, pal, you've done your big John Law bit, and
if you want, I'll go around the corner for a minute."

"Just take off," Serpico said, "and don't come back."

After he had gone, the girl said, "Frank, why did
you let him get away?"

"He was a cop," Serpico said.

"What difference does that make?"

"Jesus Christ," Serpico said. "He has a gun and I
have a gun. Did you want me to get into a shootout

with him because of a couple of bricks? Don't worry. I've got his license number, and we'll check him out."

One concrete result of *The New York Times* series on police corruption and the widely heralded Knapp Commission investigation was that Sydney Cooper had replaced Joseph McGovern as Supervising Assistant Chief Inspector in command of all the department's anticorruption units. So Serpico called Cooper and explained what had happened. Cooper exploded over the phone. "That's all you need after everything else," he yelled, "getting involved with a cop over some goddamn cobblestones!"

"Well, Chief," Serpico said as solemnly as he could, "it's not that it makes any difference to me. I'm calling on behalf of the Village Citizens Committee."

"The *what*?"

Aware that the department was completely on the defensive for the moment, and that even Cooper would react to what sounded like an organized community group, Serpico repeated, "You know, the Village Citizens Committee."

A week later Cooper grimly reported to Serpico that he could tell the "goddamn committee" that the man was indeed a cop, that he had been caught in his Staten Island backyard constructing a barbecue with the Hudson Street cobblestones, and that he was going to be brought up on departmental charges.

"Gee, Chief, I'll pass the news along right away," Serpico said. "I know the committee will really appreciate this."

But for Frank Serpico the incident had a sobering side effect. The mere fact that he had to go all the way to Cooper over something like this made it clear to him that his position in the department was untenable, and in his heart he knew that his days as a police officer were numbered.

chapter 18

In the summer of 1971, while Serpico continued to recover from his head wound, a Police Department wiretap on a Mafia phone picked up the cryptic comment that "the cop with the beard in the Village," was going to be "hit." Chief Cooper was informed, and police bodyguards were assigned to Serpico. Since Serpico had already applied for permission to go on vacation, either to Florida or one of the Caribbean islands, it was decided to allow him to leave the city, but instead of going south, as his leave papers and purchase of an airplane ticket to Florida indicated, he drove north in his Land Cruiser to Nova Scotia, and stayed in touch with Cooper by telephone.

Nothing more could be determined about the plot to kill Serpico, and on his return, although

bodyguards were again assigned to him, he asked
Cooper to remove them; he did not feel comfortable
with them around, and he insisted that he could take
care of himself.

Like everything else that had happened to him
since the day he was handed the three-hundred-dollar
payoff, it was something he would have to live with
always.

There remained a final act in the drama to be played
out. More than a year had passed since Serpico's revelations in *The New York Times* had forced Mayor
Lindsay to appoint a special, independent commission
to investigate police corruption. And for a while it
seemed that even this would come to nothing. The
commission, headed by Wall Street lawyer Whitman
Knapp, had been given just six months to do the job,
and it had predictably run out of both time and money
before it could get anywhere. That would have been
the end of the investigation had not the Department
of Justice offered a $215,000 grant to enable it to con-
tinue as part of a federal law-enforcement program.
With the money now available, the City Council, nervous about charges of a political cover-up, voted at the
last minute to extend the commission's life for another
six months.

At the same time the Knapp Commission was
struggling to survive, its legality was being challenged
in court—by the police. The Patrolmen's Benevolent
Association charged in a suit not only that the commission violated a City Charter provision outlawing
any police review board that lacked a majority of

members of the department on it, but also that the investigation might result in "great expense, harassment, and inconvenience" to policemen. The kind of harassment and inconvenience the PBA suit had in mind became clear when a separate action attacking the commission was brought by two deputy inspectors and three captains who had been ordered to answer questions about their personal finances; to do so, the five officers argued, would compel them to waive their constitutional rights "against self-incrimination."

Similar suits against the commission's subpoena powers were filed by cops from inspector down to patrolman, but they were all eventually thrown out. And when Serpico returned to the city from Nova Scotia, Chairman Knapp announced that the commission's investigative efforts and its interrogations in secret, executive session were finished and that he was ready at last to hold public hearings in October 1971.

In anticipation of these hearings, meanwhile, Commissioner Murphy continued his massive overhaul of the department, and the entire high command under his predecessor was either gone or reassigned, including First Deputy Commissioner Walsh; both he and McGovern had resigned. Among the departmental reorganization and restructuring orders that were issued almost daily, the most important was a new doctrine of "accountability," in which police commanders would be held directly responsible for the actions of their men.

Before Serpico had been gunned down, he had worked closely with the Knapp Commission's chief counsel, Michael Armstrong, a former assistant U.S. attorney; and because of the commission's limited budget and personnel, it appeared at first that he once again would have to bear the brunt of testifying about police corruption. But then a commission agent accidentally managed to trap a crooked, cold-eyed plainclothesman named William Phillips while he was extorting money from a Manhattan brothel madam, and also secretly recorded a number of additional admissions concerning graft and payoffs that Phillips had boasted about during his career as a police officer. "I'm not a fucking millionaire," Phillips modestly confided to the Knapp Commission agent. "I do OK. I live pretty good, but I'm not a millionaire."

Phillips was in a lawyer's office dickering for a ten-thousand-dollar payment to fix a fraudulent-check case when he discovered a tiny microphone and transmitter the commission agent was carrying. He was about to attack the agent when another commission operative, monitoring the conversation, rushed into the room just in time. Confronted by the evidence of his own voice describing his participation in one sordid shakedown after another, Phillips decided to make a deal. If he were granted immunity, he would turn informer. Commissioner Murphy and the various district attorneys involved then reluctantly acceded to a request by Knapp and Armstrong not to prosecute Phillips for the specific instances of corruption he confessed to; and in return he became in effect an undercover agent for the Knapp Commission, taping conversations with other policemen about graft, as well as detailing much of his own criminal past to commission investigators. And although other "rogue cops" also appeared in the hearings, Phillips became the star witness for a fascinated television audience— as he reeled off stories of pads, payoffs, and bribes amounting to millions of dollars annually and touching virtually every police division in the city. Serpico watched dispassionately; there wasn't anything Phillips said that he had not known about, and been trying to do something about, for years.

The hearings continued for nine days, and the public phase of the Knapp Commission seemed to be over. But a number of Mayor Lindsay's political enemies, spearheaded by City Councilman Matthew Troy, the Queens County Democratic leader, began charging that the Knapp investigation was not uncovering the culpability of top Police Department brass and Lindsay administration favorites, particularly Jay Kriegel. As the political pressure mounted, Whitman Knapp finally gave in and announced that there would be a new round of public hearings in which Serpico would appear before the commission.

Serpico was hesitant to testify, fearful that the whole investigation was becoming a political football, and that the central theme—police corruption, and the system that allowed it to flourish—was being lost as a result of all the infighting. He believed that the time for recriminations was long past, that it didn't matter anymore who had or had not done what, what did matter was the future, what a rookie cop coming out of the Police Academy would face. In addition, friends of Serpico's were concerned about how he might be treated in hearings that involved so many powerful municipal figures, and one of them telephoned former Attorney General Ramsey Clark, then in private practice in New York. Although Clark knew neither the caller nor Serpico, he quickly agreed to protect Serpico's interests without fee. Clark later explained, "I did it because I felt that the Police Department desperately needed men like Frank Serpico."

With Clark present as his counsel, Serpico was allowed to tell, simply and straightforwardly, the story of his lonely odyssey to combat corruption in the department. Serpico spent about three hours before the commission and the television cameras. When he had finished answering questions posed by Armstrong and other commission members, he was asked if he had anything to add. He replied that he did, and he quietly read a short statement that he had prepared the night before. Phillips had recited, better than he ever could, the morass of corruption that the Police Department had settled in—that was not Serpico's concern. His concern, the reason why he was testifying now, had always been the thousands of ordinary cops who wanted to be honest, and his statement reflected this concern:

> Through my appearance here today I hope that police officers in the future will not experience the same frustration and anxiety that I was subjected to for the past five years at the hands of my superiors because of my attempts to report corruption.
>
> I was made to feel that I had burdened them

with an unwanted task. The problem is that the atmosphere does not yet exist in which an honest police officer can act without fear of ridicule or reprisal from fellow officers.

We create an atmosphere in which the honest officer fears the dishonest officer, and not the other way around. I hope that this investigation, and any future ones, will deal with corruption at all levels within the department, and not limit themselves to cases involving individual patrolmen.

Police corruption cannot exist unless it is at least tolerated at higher levels in the department. Therefore, the most important result that can come from these hearings is a conviction by police officers, even more than the public, that the department will change.

I also believe that it is most important for superior officers to develop an attitude of respect for the average patrolman. Every patrolman is an officer and should be treated as such by his superiors.

A policeman's attitude about himself reflects in large measure the attitude of his superiors toward him. If they feel his job is important and his stature, so will he.

It is just as important for policemen to change their attitudes toward the public. A policeman's first obligation is to be responsible to the needs of the community he serves.

The department must realize that an effective, continuing relationship between the police and the public is more important than an impressive arrest record.

The system of rewards within the Police Department should be based on a policeman's overall performance with the public rather than on his ability to meet arrest quotas. And merely uncovering widespread patterns of corruption will not resolve that problem.

Basic changes in attitude and approach are vital. In order to insure this, an independent, permanent investigative body dealing with police corruption, like this commission, is essential.

After Serpico's statement—about what should be done—a dreary parade of department and city officials came before the commission, intent only on justifying their part in what they had put him through. For five years Serpico had been trying to get them one after the other to do something about the corruption permeating the force, and all they could dwell upon now was how difficult he had been to deal with.

The initial witness was cautious, funereal Cornelius J. Behan, the first ranking officer Serpico had turned to when he was about to be transferred to the 7th Division—a man who had subsequently been promoted from captain to inspector and who would shortly be appointed head of the Police Academy. Yes, Behan admitted, Serpico had called him on the phone before his assignment to the 7th Division and expressed worry about the possibility of corruption there. He said that Serpico had come to him undoubtedly because of his reputation as a deeply religious man—sort of a "father confessor" to young cops with problems—and he confirmed that as a result of Serpico's call he had contacted his neighbor Philip Sheridan, the 7th Division's administrative officer, and received assurances that Serpico had nothing to fear. He also corroborated the meetings he had had with Serpico in the winter and spring of 1967, after Serpico reported that an organized pad was operating among the division plainclothesmen. Behan said that he had three separate conferences with First Deputy Commissioner Walsh about the matter, the last in April 1967, and that the following October he had contacted Sheridan again after Serpico called in a "highly agitated state" and said that "he was not going to put up with corruption anymore, and that he was going to speak out openly against it."

This, Behan insisted, was the first time he learned that Walsh had not followed through on the information he had relayed to him. "Did you have any reason to believe that an investigation was, in fact, being conducted?" the commission chairman, Whitman Knapp, asked.

"Well, no, I had no visible signs, and I wouldn't expect to have any. But, by the same token, I had no reason not to believe it."

"Was there any reason," Knapp persisted, "why you did not inquire of the commissioner concerning the investigation?"

"Well," Behan replied, "as a captain of the police

at that time, I did not feel it my position to inquire what a commissioner was doing." Behan was not asked where this left a patrolman like Serpico, nor did he venture an opinion.

The next witness was Inspector Philip Sheridan, now retired, who had characterized the 7th Division as "clean as a hound's tooth." Sheridan grudgingly admitted that he had been familiar with many of the gambling locations that Serpico had pinpointed as paying off.

Michael Armstrong, the commission's chief counsel, asked if an investigation into those gambling spots could have taken place "whether Serpico had come forward or not."

"Yes," Sheridan said, "I suppose so."

"But in fact it had never been tried, is that right?"

"Oh, I questioned many prisoners when I was in charge of plainclothesmen. That was a customary procedure, for me to take a prisoner aside and say, 'Who are you paying?' and question him."

Sheridan reiterated that he had no previous knowledge of a division pad until Serpico reported it. He conceded that Serpico had provided enough facts "to start the investigation," but, he complained, he had refused "to wear a wire." Sheridan added somewhat plaintively that he was "never quite sure Frank Serpico trusted me." Besides portraying Serpico as an uncooperative partner in the investigation, Sheridan seemed anxious to demonstrate that he was personally free of any taint of corruption, and he concluded his testimony by noting that he was currently a high-school typing teacher at the lowest salary grade and thus was going to bill the Knapp Commission for loss of income resulting from his appearance before it.

The theme that the 7th Division investigation had been a great success *despite* Serpico's balky attitude— evidenced by his refusal to wear a hidden wire recorder—was quickly taken up by Jules Sachson, now Deputy Chief Inspector, who once said to Serpico that he had dropped an unwanted "hot potato in our laps," and by Bronx District Attorney Burton Roberts, who told the commission that the investigation "could have accomplished" even more if Serpico had not been so uncooperative. "This man didn't want to wear a wire," Roberts growled. "This man didn't want to testify. This man didn't want to go into the grand jury." But under questioning from members of the commission, Roberts, his tone more subdued, finally agreed that without Serpico "there would have been no police investigation" and "no grand jury investigation."

Then former Supervising Assistant Chief Inspector Joseph McGovern, whose favorite meeting place with Serpico was the Fulton Fish Market, presented himself as a friendly amateur psychologist, the kind a fellow in trouble needed. Serpico, he said, was a man who required "some guidance and assistance . . . and this was the role that I feel I played." It was why, McGovern explained, he did not do anything about the matter of Captain Foran and the three-hundred-dollar envelope that Serpico had told him about until just before *The New York Times* story on police corruption was published. "I remember," McGovern said, "that Frank had a lot of inward pressure on him," and he did not want to take any steps that "added to [Serpico's] burden." McGovern maintained that he could not recall Serpico's ever telling him about the recording he had made of the cops extorting money from his brother, and he added some intriguing insights into the way First Deputy Commissioner John Walsh operated. Walsh was a man whom one did not "ask many questions," according to McGovern. If one wanted his opinion on a particular subject, no matter how complicated, and he simply answered, "yes"—well, that was it. "When he says 'yes' that's the end of the discussion," McGovern observed, his voice still awed.

When John Walsh himself appeared before the commission, it was difficult to believe that this was the man whose very name struck terror in the heart of each cop in the department. In strained, bumbling replies to queries from chief counsel Armstrong and from commission members, Walsh tried to explain away the great mystery of why he had allowed Serpico to stay in limbo in the 7th Division from April 1967, when Behan last saw him, until October, when Serpico contacted Behan again and Behan in turn called Sheridan. Walsh said that he had simply forgotten about it.

He did not recall the first two of the three meetings Behan said he had with him, Walsh testified, but he did remember the third one in April. "I said I would meet with Serpico," Walsh told the commission. "Unfortunately that meeting never took place. I expected the meeting to be arranged by Inspector Behan. When it was not arranged, the incident left my mind."

One could not tell whether during the last years of

the enormous solitary power Walsh had wielded in the department—remote, confiding in no one, keeping tabs on everybody—his mind had actually been as uncertain as it now appeared, or whether this was simply another Machiavellian bit of evasiveness on his part. Chairman Knapp tried to find out without success.

"Why the hell did you wait six months before checking with Behan?" he demanded in an uncharacteristic outburst.

Walsh stared back blankly at him—and did not reply.

Next David Durk, now a sergeant, confirmed his presence at the meetings Serpico had with Captain Foran, Jay Kriegel, and Investigation Commissioner Arnold Fraiman. He added that in a subsequent encounter, Fraiman had dismissed Serpico as a "psycho." Then, with an impassioned plea to the Knapp Commission and the public to bring an end to police corruption, his voice breaking, Durk rushed from the conference hall where the commission hearings were being held.

The horror story continued with the swearing in of Arnold Fraiman, now a New York State Supreme Court Justice. Fraiman acknowledged that he had met with Serpico and Durk, but his recollections of the meeting, he acknowledged, were "hazy." He denied ever calling Serpico a "psycho," but confessed that he thought Serpico in his appearance was an "unusual individual." Fraiman insisted that Serpico spoke only in "general terms" about corruption in the 7th Division. At one point during his testimony he claimed, "My immediate reaction was that I did not feel the information was that valuable." At another point he contradicted this and said, "I did not question the validity of the information. I may have thought it was difficult to prove."

"I find it hard to understand," chief counsel Armstrong acidly remarked, "why this important and serious allegation was not referred to *someone*."

"I normally would have done so, but he [Serpico] did not want to be identified."

"It is difficult," Armstrong persisted, "to understand why this information was just dropped," and Chairman Knapp broke in to inquire if Fraiman felt he might have committed "an error in judgment."

"If I did so," Fraiman archly replied, "it was not the only error in my three years in the Department of Investigation," as though that ended the matter.

The most incredible performance of all was given by former Police Commissioner Howard Leary. Nobody ever told him anything, he said, suggesting that things might have been different if they had. The first he had ever heard of Serpico or the scandal in the 7th Division was on October 9, 1967, when John Walsh mentioned it to him after meeting with McGovern and the officers from the Bronx. After that, Leary continued, he simply left the problem in the capable hands of Walsh. In a press conference following his appearance before the commission, Leary, who had privately concurred with Foran in calling Serpico a "psycho" and who had publicly described him as a "disgruntled" cop, blandly told reporters that when he had seen the "anguish and frustration" of Serpico on television before the commission, "it was a horrible thing."

The last witness to testify before the Knapp Commission was Jay Kriegel, now Mayor Lindsay's chief of staff. Kriegel confirmed that Durk, whom he knew well, had brought Serpico to his City Hall office in April 1967. He said that he had then given Mayor Lindsay a general briefing about Serpico's allegations of corruption in the 7th Division, but that Durk's ground rules for the meeting—that he and Serpico had to be guaranteed anonymity—had "handcuffed" him in pursuing the matter further. "I told the Mayor that I had met with a police officer through a mutual friend; that he had talked to me about the problem of corruption, and that they had suggested pulling together a group of officers, like Frank Serpico, with similar stories, for the Mayor to hear this information firsthand."

According to Kriegel, the Mayor rejected the proposal because of the secrecy demand tied to it. Michael Armstrong inquired as to what the Mayor's response had been to the corruption charges. "He asked me if the allegations had been reported to the department," Kriegel said, "and I told him they had."

"And did you also report to him that the police officers were dissatisfied with the pace of the investigation?"

"No," Kriegel said, "I did not."

Then Armstrong dropped something of a bombshell. Had not Kriegel under oath in executive session said he told the Mayor that Serpico and Durk were "not satisfied" with the progress of the department's investigation into Armstrong's charges?

Kriegel replied that after reading the record of his closed-door testimony, he had "talked at some length to the Mayor about that incident, and I am clear now that following the meeting with Frank and David, that I did not report to him that allegation."

For a time the Manhattan district attorney's office considered indicting Kriegel for perjury. That it did not was primarily the fault of the Knapp Commission's interrogation of him in executive session earlier in the year. While the commission, smarting under the charges of a political whitewash, was now taking a tough stance against the Lindsay administration, it had done just the opposite before. In the two-hundred-page, double-spaced transcript of Kriegel's secret testimony, much of it covering the information provided by Serpico, barely more than a page dealt with what Kriegel had or had not passed on to Mayor Lindsay. According to that testimony, Kriegel stated to the Mayor that "police officers" had charged that the department was not following through on their allegations of internal corruption. Kriegel was quoted as saying, "I made recommendations to the Mayor on the general problem of corruption, including that as one indication of the problem." But, typically, none of Kriegel's executive-session interrogators pushed him as to what he meant by "that," and he was able to claim that the "that" simply meant that police officers had come to him and did not refer to any specific complaint they had. If his answers were not "thorough," Kriegel told the D.A.'s office, it was because the commission's questions were not. The best reflection of the commission's attitude at the time was Chairman Whitman Knapp's rhetorical question to a reporter who asked how the commission was going to handle Lindsay in its investigation. How, Knapp blurted back, can you investigate the man who appointed you?

Possibly the truest explanation of what really happened lay in Mayor Lindsay's response to a question from a student at the University of Pennsylvania when he spoke there shortly after the story on police corruption broke in *The Times*. Why, the student wanted to know, had it taken a newspaper exposé "to move the city"?

"If you've had as long and as delicate a relationship with the thirty-two-thousand-member Police Department as I have had," Lindsay said, "you might understand."

So it was over, and throughout the winter of 1972 Frank Serpico remained on sick leave, recovering from the wound in his head that had nearly cost him his life. New X rays showed that while some of the bullet fragments had moved, the critical fragment, the one nudging his carotid artery, had not. All through the winter he also debated his own future as a cop, sensing the underlying hatred of the police establishment toward him after what he had done, knowing that he would forever remain someone apart. He wondered how much longer he could endure such a pariah-like existence.

In May a new scandal hit the Police Department. Twenty plainclothesmen, one policewoman, and three sergeants in the 13th Division in Brooklyn—the division where Armstrong had been given the three-hundred-dollar envelope—were indicted for taking a quarter of a million dollars a year in bribes from gamblers linked to the Mafia. This scandal was disclosed not because another honest cop like Serpico had come forward, but because another dishonest one had been caught. Following the Knapp Commission hearings, the department had adopted a new policy of offering immunity in certain cases to police officers caught in corruption if they would cooperate in uncovering more corruption, and it had led to the 13th Division indictments. Not long after, still another scandal erupted, this time involving members of the elite Special Investigating Unit of the Narcotics Bureau—the unit to which Serpico, when he became a narcotics cop, had been instructed to turn over anything "big" that he stumbled across.

When, Serpico thought, would it all end? But the thing that finally made up his mind about whether or not to remain a cop came in the spring of 1972, when the Police Department announced that he would be given its highest award, the Medal of Honor—not for his courage in reporting corruption, but because, as he put it, he "was stupid enough to have been shot in the face." And so he decided it was time to quit, and now neither Sydney Cooper nor Paul Delise, the two men he most admired in the department, was in a position to dissuade him. Cooper had already told him that he was going to retire, and Delise was getting ready to do the same.

On June 13, nearly five years after Serpico had first reported him, Captain Philip Foran finally went on departmental trial for making false statements about his advice to Serpico regarding the envelope and the three hundred dollars. Both Serpico and Durk testified against him, and Foran, who continued to deny their charges, was at last found guilty. His punishment, however, was simply a slap on the wrist— suspension without pay for thirty days. Captain Foran appealed the charges against him on the grounds that they were "arbitrary, capricious, and unlawful." His appeal claimed that the Police Commissioner had exceeded his authority, since a grand jury looking into the matter in 1970 voted "no true bill," and furthermore that the statute of limitations had run out regarding his meeting with Serpico and Durk. He also charged that the departmental trial was "politically motivated" and prompted by "undue publicity."

By then Serpico had officially ended his career as a policeman. He left the force with a disability pension and a gun permit for his Browning automatic. He decided that he would leave the country for a while to sort out his shattered life. At the age of thirty-six, he figured that he could still contribute something— somewhere, somehow.

As he was walking out of Police Headquarters with his papers, another cop came up to him, as many had done following the Knapp Commission hearings, and said with great seriousness, "Gee, Frank, do you think you really changed anything? Do you think things are going to be different?"

"I don't know," Serpico replied. "It's not up to me anymore. I only did what I had to do."

Serpico's leaving was a tragedy for the city, for the Police Department, and for himself. All Frank Serpico ever wanted was to be a good cop. Perhaps that was the trouble; he had wanted it too much.

afterword

"Only a fool, fixed in his folly, thinks that he can turn the wheel whereon he turns."
—T.S. Eliot

"Perhaps nobody yet has been truthful enough about what 'truthfulness' is."
—Fredrich Nietzsche

Seated in front of my cabin in the woods, I feel the sun hot on my face. The sky is clear except for a few small clouds. Though it is already late October, some flowers have defied the evening frosts. A train whistle echoes in the distance; the river reflects the sun like a giant silver mirror. This is my Walden. My healing place.

Twenty-five years have passed since the night of February 3, 1971. But the events of that evening remain fixed in my mind, like the lead bullet fragments I carry fixed in my head. For most of those 25 years, I traveled Europe and the U.S. My many encounters with people, especially Americans, have led me to believe there exists in the minds of many, a resistance to change.

I left New York with my Old English sheepdog, Alfie, on a transatlantic luxury liner, first class. The book *Serpico* was not yet released, nor the movie made, but my picture had appeared in *Newsweek* magazine. The trip was a wonderful opportunity for me to get away and travel incognito, as I desired. I dressed casually in my old clothes, costumes I had become accustomed to wearing as an NYPD narc, living in Greenwich Village. I still displayed a wristful of assorted bangles and bracelets, two earrings, and the hair and beard of Rasputin. However, my wardrobe did include a 3-piece custom-made suit from Milan, a couple of handmade ties, and a pair of soft leather boots. On more than one occasion, I was approached by crew members and commanded to "go back to your class." To my amusement and annoyance, they assumed I was traveling tourist class, not cabin class, and certainly not first class. I thought to myself, *they would never approach me in this fashion if I wore my 3-piece suit.* Unable to place my demeanor with my dress code, the ship's steward one day asked me to meet him on the sun deck at four o'clock. Curious, I attended to hear him lay out how he and I could throw the coming evening's bingo game and split the cash jackpot. I could only look at him and smile. The NYPD did not have a monopoly on corruption, after all.

On arriving in Europe, Alfie and I traveled by car from Naples to Zurich. Switzerland was like a fairytale; even the police were a pleasure to deal with. They actually helped me find lodging. Once or twice, I received warnings for traffic infractions but no tickets. In fact, one morning I found a parking ticket on my windshield; upon presenting the ticket with an explanation of not understanding the traffic sign, I watched dumbfounded as the officer tore the ticket in half, saying in French, "At your service." I felt like Alice in Wonderland. This all happened before the French edition of *Serpico* was released. The officer didn't know me from Adam; I was just another tourist. I immediately flashed back to the police car pound in New York; a tourist with a thick French accent had come to pick up his impounded vehicle. Unfortunately, he was a few dollars short of the towing fee. He offered the officer on duty a brand new shirt to make up the difference. The cop looked at the shirt collar and said, "Sorry, not my size." I lent the tourist the rest of the money; now in Switzerland, the courtesy was being returned.

In the French Alps, I rented a remote nineteenth-century cold-water chalet. I was still under medication, three times a day, for the bullet fragments lodged in my skull (I was told it would be for the rest of my life). Movie director John Avildsen (*Rocky, The Karate Kid*) and writer Norman Wexler came to visit and together we worked on the screenplay for the movie. Avildsen was eventually replaced by Sidney Lumett. Later, I was asked to go back to New York to help out on the movie. As it turned out, it wasn't worth the hassle, as I soon realized I wasn't really wanted there. But the time spent with Al Pacino was very pleasant. I found him to be a man of character.

One night in Holland I witnessed a burglary in progress on a boat anchored across the canal outside my window. I called the police and by the time Alfie and I arrived at the scene, the police were in the process of apprehending the suspect as he emerged from the boat, his arms full of booty. I had never witnessed a suspect handled so professionally and humanely. I heard myself actually telling the police, "I don't want to get involved," a statement I had so often heard from witnesses at crime scenes. The next morning, two Dutch detectives were at my door. The police commissioner requested my presence at headquarters. I was welcomed with a big smile and a handshake and asked would I please autograph his hardcover English edition of *Serpico*. I signed the inscription, "It's always a pleasure to shake the hand of an officer who wants to shake my hand."

Some time later, I was a guest on a New York radio program. A New York City police officer called in to say, "You know, Frank, the day you testified before the Knapp Commission was a dark day for every cop in New York City. When I went home that night, I couldn't face my wife and kids." "Why?" I inquired, "What did you do wrong?" "Nuttin'," he replied. "Then why didn't you come out and back me up?" I asked. Without hesitation he shot back, "What! And be an outcast like you?" A very sad attitude, one that I believe reflects that of most honest officers. It is my opinion that there have never been any real department incentives for cops to be honest, from the police academy up through the ranks to the police commissioner.

An excuse often proffered by police officers for their corruption is that seeing such enormous sums of money, on their salary, is too much temptation. If this form of logic held true, bank tellers would be taking money home by the bagful. Unfortunately, the real culprits of my story were never brought to task. They went on to become judges, politicians, commissioners, and university professors. Men of base character are elected to the highest offices of the land. In the final analysis, the people get the police they deserve.

I eventually returned to Holland and bought an old thatch-roofed farmhouse. I settled down with a wife and two children, a couple of horses, some sheep, and an assortment of other animals and fowl; plus Alfie, my sheepdog, now had two new canine companions. We had all the material things people think they need for happiness: life was good. By some quirk I was assigned the local police station's former phone number. People kept calling looking for the police, and at first, I thought it was a joke. In town I realized that the authorities had failed to remove the old police number from the public phone booths.

My wife died from cancer at the age of 29. Afterwards, to occupy my mind, I started traveling again. Most of my encounters with police officers in Europe, the States, and other countries have all been quite pleasant, and I am grateful for the continued support and cooperation that they have afforded me. After almost ten years abroad, it was time to return to America, through Canada, down to Mexico, and back up to New York, my Brooklyn apartment, and my cabin in the woods.

All in all, the journey that began on the night of February 3, 1971 has thus far been enlightening, with all its joys and sorrows. Although I had not realized it at the time, that evening proved to be the major turning point in my life.

It wasn't until a few years ago while listening to a radio program about people who had near-death experiences that I realized what actually happened to me on that fateful night. Upon being shot in the head, I saw parts of my life replayed in slow motion, all in a matter of minutes. I had a vision and heard a voice speak like thunder. It said, "It is all a lie." In all the deaths I have witnessed, I always found myself looking down on the lifeless bodies. Now I was looking down on a body in a pool of blood wearing my clothes. It was like, "Hmmm, he's got my clothes on." Then, almost as fast as it happened, I was back in my body looking up at the hallway ceiling. This experience has since caused me to seriously reflect on the issues of life, death, and religion. Albert Einstein summed up this feeling well on his death bed: *I am just Energy and I am Indestructible.*

Of even greater personal significance are my father's last words to me: "Be careful son, there are a lot of bad people out there."

"I know Pa, but there are a lot of good people, too," I replied.

"Yes, but you don't have to worry about them."

—Frank Serpico
Upstate New York
Fall 1996

BOOKS BY PETER MAAS

THE TERRIBLE HOURS: *The Greatest Submarine Rescue in History*
ISBN 0-06-093277-5 (paperback)
On the eve of World War II, the Squalus, America's newest submarine, plunged into the North Atlantic. Miraculously, thirty-three crew members still survived—their ultimate fate depending upon one man, U.S. Navy officer Charles "Swede" Momsen.
"A suspenseful tale of terror, courage, heroism, and American military genius."
—Tom Brokaw

SERPICO
ISBN 0-06-073818-9 (paperback)
For years a culture of corruption had pervaded the New York City Police Department. Into this maelstrom came a man who broke the mold: a working class, Brooklyn-born, Italian cop with long hair, a beard, and a taste for opera and ballet. Most of all, Frank Serpico was a man who couldn't be silenced—or bought.
"[A] raw and moving portrait." *—Chicago Sun-Times*

UNDERBOSS: *Sammy the Bull Gravano's Story of Life in the Mafia*
ISBN 0-06-093096-9 (paperback) • ISBN 0-06-109664-4 (mass market paperback)
In 1992, the highest-ranking member of the Mafia in America ever to defect broke his oath of silence. Gravano's story brings us into the innermost sanctums of the Cosa Nostra—a secret underworld of power, lust, greed, betrayal, and deception, with the specter of violent death always waiting in the wings.
"An absorbing, intimate, alluring tale of power, greed, and Mob intrigue." *—People*

FATHER AND SON
ISBN 0-06-100020-5 (mass market paperback)
The powerful and moving story of Michael McGuire, a successful New York executive and lonely widower who is suddenly drawn into an IRA gun-running plot and must immerse himself in a violent and twisted web of intrigue in order to save his only son.
"A powerfully tragic story." *—New York Times*

THE VALACHI PAPERS
ISBN 0-06-050742-X (paperback)
When Joe Valachi decided to violate the Cosa Nostra oath of silence, he told everything. His recollection of over 40 years of associations traces the growth and development of the Mafia in the United States, revealing its membership, operations, and methods.
"Littered with bodies and unsolved crimes, betrayals and beatings, oaths, ritual, and revenge." *—Newsweek*